ANTHONY KENNY

# What is Faith?

# What is Faith?

## Essays in the Philosophy of Religion

Anthony Kenny

Oxford New York
OXFORD UNIVERSITY PRESS

*Oxford University Press, Walton Street, Oxford* OX2 6DP
*Oxford New York Toronto*
*Delhi Bombay Calcutta Madras Karachi*
*Kuala Lumpur Singapore Hong Kong Tokyo*
*Nairobi Dar es Salaam Cape Town*
*Melbourne Auckland Madrid*

*and associated companies in*
*Berlin Ibadan*

*Oxford is a trade mark of Oxford University Press*

*Published in the United States by*
*Oxford University Press Inc., New York*



Faith and Reason *first published by Columbia University Press 1983*
What is Faith? *first published as an Oxford University Press paperback 1992*



*British Library Cataloguing in Publication Data*
*Data available*
*ISBN 0-19-283067-8*

*Library of Congress Cataloging in Publication Data*
*Kenny, Anthony John Patrick.*
*What is faith?: essays in the philosophy of religion/Anthony Kenny. p. cm.*
*1. Faith and reason. 2. Religion—Philosophy. I. Title.*
*210—dc20 BL151.K43 1992 91-43539*
*ISBN 0-19-283067-8*

3 5 7 9 10 8 6 4 2

*Printed in Great Britain by*
*Biddles Ltd.*
*Guildford and King's Lynn*

# CONTENTS

# PREFACE

This first part of this book consists of the four lectures which I gave in April 1982 as the twenty-second series of Bampton Lectures in America. These lectures were published in 1983 by Columbia University Press under the title *Faith and Reason*.

Oxford University Press, in agreeing to bring out a paperback edition of these lectures, suggested that I should include along with them some more recent lectures, given to different audiences, which address connected topics and develop similar themes.

Accordingly, the second part of the book contains the following papers. 5. 'Is Natural Theology Possible?', an adaptation of a lecture delivered at the December 1989 session of the American Philosophical Association at Atlanta, Georgia, and not previously published; 6. 'The Argument from Design and the Problem of Evil', a lecture delivered to a symposium in Rome in 1987 and published in 1988 in vol. lvi of *Archivio di Filosofia* entitled *Teodicea Oggi?*; 7. 'John Henry Newman on the Justification of Faith', an abbreviation of a lecture on Newman as a philosopher of religion, delivered in a series of centenary celebration lectures in Oxford in 1990 and published in *Newman, a Man for our Time*, ed. David Brown (SPCK, 1990); 8. 'Anselm on the Conceivability of God', a lecture delivered to a symposium in Rome in 1989 and published in 1990 in the series *Biblioteca dell' Archivio di Filosofia* under the title *L'argomento ontologico*.

I am grateful to all those under whose auspices these lectures were given, and to the audiences who attended the lectures and made comments which enabled me to improve them with a view to publication.

# PART ONE
# Faith and Reason

# 1 The Virtue of Reason

The topic of these lectures is one of the most central issues in the philosophy of religion: the question whether belief in God, and faith in a divine word, is a reasonable or rational state of mind. In these lectures I shall offer an answer to the question: is faith rational? I shall do so by trying to define more precisely what faith is and what reason is.

Surprisingly, I have found it much more difficult to define what reason is than to define what faith is; and so I shall spend the first two lectures considering the nature of reason. Only in the last two will I consider the nature of belief in God, or faith. In the first lecture I will discuss the most popular philosophical account of reason or rationality, and I will reject it; in the second I will try to offer an improved account in its place. Then in the third lecture I will ask whether belief in the existence of God is compatible with rationality as I have defined it; and in the final lecture I will ask the same question about another kind of belief, faith in a divine revelation. My first topic is the nature of rationality: the virtue of reason. But I must first say just a few words about what I mean by 'belief in God'.

There are a number of different states of mind which may be described as 'belief in God'. We may distinguish between three senses of that expression.

1. Belief that there is a God, that God exists.
2. Belief in a doctrine on the word of God, as revealed by God.
3. Belief in God as trust in God and commitment to Him.

Belief in God in the first of these senses is belief in the truth of the proposition 'God exists'. It is a belief which might be reached in many different ways: one might believe it on the basis of a proof, or because one was taught to believe it in Sunday School or because a Godless world is too horrible to contemplate. It is a belief which may be held with varying degrees of conviction and which may find expression in many different types of behaviour, from a mere inclination to answer 'yes' to the opinion pollster's question 'Do you believe that there is a God' to a life devoted to what one believes to be the divine service.

Belief in God in the second sense is more than the mere belief that God exists. It is belief in some different proposition on the basis that it has been revealed, or vouched for, by God. It is not so much believing in God, as believing God: taking something for true on the word of God. Thus, one may believe that Jesus Christ will return and judge the world, believing that this is a truth which has been revealed by God, which God has given his word for. Again, one may believe that the children of Israel have a unique and permanent destiny, on the basis that this is a promise which God has made. Belief of this kind is not simply belief but faith, the faith for which Abraham was praised and for which Paul contended.

Belief in the third sense is more than the mere intellectual commitment to the truth of certain propositions as revealed by God. It involves a resolution to act upon these propositions: a commitment of oneself to the revealed purposes of God, a trust in His enabling one to enact them in one's life. It can be described not only as belief in God, but also as love of God: in its fullest manifestation, as the love of God above all things.

At the time of the Reformation much ink was spilt, and much blood was shed, about the relationship between belief in the second sense and belief in the third sense. For the Catholic tradition, faith consisted essentially in the intellectual assent to doctrines as revealed by God. Such faith was one of the three key 'theological virtues' of faith, hope, and charity; it was a virtue, and therefore a valuable and praiseworthy thing, whether or not it was accompanied with the loving commitment to God, which was charity. If so accompanied, it was living faith; if not it was dead faith. For the Protestant Reformation the only faith worthy of the name, the faith eulogized in the Bible, was trust in and commitment to the saving purposes of God.

I shall not be concerned here with this question about the relationship between belief in the second sense and belief in the third sense. It is an important question, but it is a question of theology and of history; and in these lectures I shall be concerned rather with philosophy of religion, since I am a philosopher and not a theologian or a historian. As a philosopher, I shall be discussing the nature of belief in God considered in the first and second senses.

The question I shall consider is the most important question in the philosophy of religion. It is the question whether belief in God—considered in either of these two senses—is or is not rational, is or is not worthy of a reasonable human being. I shall thus be engaged in considering the relationship between Faith and Reason.

From the outset, however, it is important to guard against possible misunderstanding. Faith and Reason are sometimes presented as two contrasting sources of information about religious matters: thus a Catholic theologian might maintain that there are some truths about God (e.g. that he is omnipotent) which can be discovered by unaided reason, while there are others (e.g. that there are three persons in one God) which are unattainable without the grace of faith. I shall not be developing a contrast of this kind; rather, I shall be asking a question about both belief in the existence of an omnipotent God and about faith in the doctrine of the Trinity: namely, the question whether either belief is reasonable or irrational.

Reason, besides being contrasted with faith, is sometimes contrasted with understanding or intuition: some truths, we are told, are seen to be true as soon as they are understood; others are only discovered by more or less complicated, lengthy, and arduous processes of ratiocination. This contrast too, important though it is, is not of immediate relevance to the question I shall address of the reasonableness of belief in God. That question is concerned with reasonableness or rationality in a particular sense which I shall now attempt to make more precise.

It is important for human beings to strike the right balance in belief. One can err by believing too much or believing too little. The person who believes too much suffers from the vice of credulity or gullibility; the person who believes too little is guilty of excessive incredulity or scepticism. If you believe too much, your mind will be cluttered with many falsehoods; if you believe too little you will be deprived of much valuable information. There is no universally accepted name for the virtue which stands in the middle between the two vices of credulity and scepticism: a name which is sometimes used, and which is as good as any, is 'rationality'. It is in this sense that I am using the word *rational* when I inquire whether religious belief is rational. The rational human being is the person who possesses the virtue that is in contrast with each of the opposing vices of credulity and scepticism.

It was Aristotle who first made familiar the idea that moral virtues stand in a mean: that is to say, each virtue is flanked by two opposing vices, and each virtue is a disposition to have or do the right amount of something of which there can be vicious excess or defect. Thus the virtue of courage stands in the middle between cowardice and rashness: the courageous man has the right amount of fear, the coward too much and the rash person too little. Again, the liberal man spends the right amount of money on suitable objects: the miser spends too little, and the spendthrift too much. Adopting Aristotle's apparatus, I can describe rationality as a mean between scepticism and credulity, as the virtue which determines the mean in matters of belief.

Aristotle himself did not identify any virtue which had belief as its field of operation. Indeed, in the *Nicomachean Ethics* he says that intellectual virtues are not concerned with a mean as the moral virtues are. In the *Eudemean Ethics,* however, he says that the intellectual virtue of *phronesis* or wisdom is a mean between cunning and folly. A practical intellectual virtue, then, like wisdom, may be regarded as a mean; but when Aristotle comes to treat of mental states concerned with theory, in the book on intellectual virtue which is common to both his ethical treatises, he does not invoke the doctrine of the mean at all. This is because he concentrates his attention on those mental states which have only truths as their objects, such as knowledge and understanding. Because truth is the good of the intellect, and because whatever is known is true, there cannot be too much knowledge and therefore there is no need to try to identify a virtue whose role is to see that one has just the right amount of knowledge. But belief, or opinion, as Aristotle himself remarks, is a state of mind which may be either true or false. If something is false, then I do not know it, however much I may think I do; but a belief of mine may be false and yet remain a perfectly genuine belief. There is room, then, for a virtue which determines the mean, the right amount, of belief; and it is a gap in Aristotle's system that he does not consider this virtue.

What general account can we give of the virtue of rationality? According to Aristotle an act of any virtue must be in accordance with a 'correct thought', an *orthos logos*. The *orthos logos* involves two things: a correct general appreciation of the nature of the virtue in question, and the application of this to the circumstances of the particular case. Only the wisdom of the individual, born of good will and experience, can determine the action to be done in the light of the circumstances; but the nature of the general criterion for a particular virtue is something about which there can be theoretical discussion, and about which philosophers may disagree, as they disagree, for instance, whether justice involves giving to each according to his deserts or to each according to his needs. So too with the virtue of rationality in belief: the individual's integrity, experience, and wisdom will determine what is to be believed and what is not to be believed in the individual instance; but general criteria for rationality in belief are something which the philosopher can and should try to formulate in the abstract. In so doing he will be, in Aristotelian terms, making explicit the general premise in the *orthos logos* of the virtue of rationality.

It is not a simple matter to specify the criterion of rationality of belief. It is easy to formulate criteria for the credulous person: e.g. 'believe everything you are told'. It is easy to formulate criteria for the sceptic: e.g. 'believe only what you see with your own eyes'. Anyone who followed either of these prescriptions would end up believing too much or too little. How do we formulate the criterion for the rational believer who stands between the vices of excess and defect?

It might be suggested that one should believe only what one knows. Perhaps, indeed, Aristotle's failure to introduce a virtue of belief beside the virtue of knowledge could be taken as a commendation of this criterion. But such a criterion would undoubtedly err on the side of scepticism, and would cut the believer off from many truths to which he has access even though he cannot achieve the certainty about them necessary for knowledge.

Shall we say, then, that the criterion of rationality in believing is that the believer should accept only true beliefs? Even this is excessively restrictive: there seems no doubt that in the appropriate circumstances it may be reasonable to believe a proposition that is, in fact, false. It is reasonable for laymen to believe what they are told unanimously by the experts in a particular science; but from time to time the unanimous opinion of experts turns out to be wrong.

Many philosophers have proposed that the test of rationality in belief is that belief should be in proportion to evidence. Thus John Locke wrote that the mark of a rational person was 'the not entertaining any proposition with greater assurance than the proofs it is built upon will warrant'. In our own time W .V. O. Quine and J. S. Ullian in *The Web of Belief* have written: 'Insofar as we are rational in our beliefs . . . the intensity of belief will tend to correspond to the firmness of the available evidence. Insofar as we are rational, we will drop a belief when we have tried in vain to find evidence for it.' Similar quotations could be collected from many philosophers who came between Locke and Quine.

If this is the correct criterion of right belief, then it is clear what we must do in order to decide whether belief in God is rational. We have to discover whether the evidence for the existence of God is sufficient to warrant the degree of assent characteristic of a believer. In a later lecture, I shall consider the question whether there is evidence for the existence of God, and if so, what it may be. But in the present lecture I wish rather to question whether it is correct to say, as so many distinguished philosophers have said, that the mark of rationality is the proportioning of one's belief to the evidence. I want to ask whether a person can believe something rationally without having evidence for that belief. Can there be rational beliefs for which there is no evidence at all?

To answer this, we must inquire what kind of thing *evidence* is. A proposition may itself be evident, or it may be something for which other propositions provide evidence. Propositions which are themselves evident may either be self-evident (as that $2 + 2 = 4$) or be evident to the senses (as that it is snowing as I write this). Such evident propositions may provide evidence for propositions which are not in themselves evidence: as the occurrence of muddy marks on the carpet (evident to the senses) may provide evidence that the children have returned from school (not yet evident to sight or hearing); or as a long multiplication may prove, from self-evident truths like $2 \times 2 = 4$ and $3 \times 4 = 12$, such non-self-evident truths as $34 \times 13 = 442$.

Things which are evident may be said to be believed without evidence. Certainly there is no other proposition put forward as evidence for them. It is a mistake, I believe, to regard propositions which are evident to the senses as being known by inference from propositions about appearances: my knowledge that it is snowing is not a deduction from propositions about snowlike visual impressions. It is equally a mistake to regard self-evident propositions as resting on themselves as evidence: nothing can provide evidence for itself, any more than a witness can corroborate his own story. So evident propositions are believed without evidence.

It is clearly rational to believe what is self-evident, or evident to the senses. To this extent, therefore, it is rational sometimes to accept propositions without evidence. Those who say that rational belief must be proportioned to evidence must therefore to this extent modify their position. Most would be happy to accept the modification as compatible with what they really wished to maintain. A belief is rational if it is in a proposition which is self-evident, evident to the senses, or is in proportion to the evidence provided for it by such propositions. Rational belief, then, will either itself be evident, or be based directly or indirectly on what is evident.

Many philosophers, both theists and atheists, have accepted this criterion for the rationality of belief. Many theists have regarded it as appropriate to apply this test of rationality to belief in the existence of God. Naturally, being theists, they maintained that when the test was applied to the belief it passed the test. For Aquinas belief in God could be shown to be rational because the existence of God followed deductively from propositions which were evident to the senses, such as 'some things move' and 'some things, e.g. pepper and ginger, are hot'. For Descartes the belief was rational because it was self-evident, or rather it could be made so by careful meditation on the concept of God, meditation articulated in the celebrated ontological argument for God's existence. Atheist philosophers, on the other hand, such as Bertrand Russell, have rejected the existence of God on the grounds that there was insufficient evidence for it. Common to theists like Aquinas and Descartes, and to an atheist like Russell, is the premise that the rationality of a belief must be tested by its relationship to a set of basic propositions which form the foundations of knowledge. This common belief has been given the apt name 'foundationalism' by Professor Alvin Plantinga. I am greatly indebted to Plantinga's work, and I intend to develop my account of rationality in belief by expounding and criticizing the discussion of foundationalism in a number of recent papers of Plantinga, in particular his essay 'Is Belief in God Rational?'

Plantinga takes as a spokesman for foundationalism W. K. Clifford, the author of the famous essay 'The Ethics of Belief'. Clifford sums up his view by saying 'It is wrong always, everywhere, and for anyone to believe anything upon insufficient evidence.' He makes clear that he believes that anyone who believes in the existence of God does so on insufficient evidence and therefore sins against the ethics of belief; is guilty, in our terms, of the vice of credulity.

The essence of the Cliffordian position, according to Plantinga, is that there is a set of propositions *F* such that my belief in God is rational if and only if it is evident with respect to *F*. Let us call the assemblage of beliefs a person holds, together with the various logical and epistemic relations that hold among them, that person's *noetic structure*. Then the set *F* will constitute the foundations of the noetic structure; and for each person *S* a proposition *p* will be rationally acceptable for *S* only if *p* is evident with respect to *F*.

Now is belief in God evident with respect to *F*? May not, Plantinga asks, belief in God itself be a member of *F*, and be itself part of the foundations of a rational noetic structure? No, says the classical foundationalist: the only propositions which properly belong in the foundations are those which are self-evident or evident to the senses. (According to some foundationalists, whom Plantinga discusses at length, even propositions which are evident to the senses, such as 'it is snowing' do not properly belong in the foundations: their place should be taken by incorrigible propositions concerning immediate experience. But this is an unnecessary complication based on a misunderstanding of the nature of propositions which are manifest to the senses, and we need not pursue it.)

We must ask, what is self-evidence? Plantinga makes two points: self-evidence is relative to persons, so that what is self-evident to one person need not be self-evident to another; and self-evidence has two components, one epistemological and the other phenomenological. First, the epistemic component: a proposition is self-evident only if it is known immediately. Second, the phenomenological component: a self-evident proposition has about it 'a kind of luminous aura or glow when you bring it to mind or consider it': what Descartes called 'clarity' and Locke called an 'evident lustre'.

The question arises: how does one know whether a thing is self-evident? For not everything that seems self-evident turns out to be so. It seems self evident, for instance, that some classes are members of themselves: the class of classes is itself a class. It seems self-evident that others are not: the class of men is not a man. It seems self-evident, further, that every class either is or is not a member of itself, so that there is a class of classes which are not members of themselves. But Russell has shown that this, so far from being self-evident, is not true and does not even make sense. The appearance of self-evidence, therefore, is no guarantee of self-evidence.

The foundationalist, therefore, if he is to justify his appeal to self-evidence, must accept some such proposition as 'whatever seems self-evident is very likely true'. But such a proposition is neither self-evident, nor evident to the senses: so its acceptance violates the foundationalist's canon of rationality, if he accepts it, as he does, without reason as basic. Moreover the foundationalist's canon itself, that nothing is to be accepted as basic unless it is self-evident, or manifest to the senses (or, in the other version of foundationalism, incorrigible), is itself something which he accepts as basic; so he is hoist with his own petard, for it is itself neither self-evident nor manifest to the senses nor incorrigible. Why should we accept such a canon?

The answer [Plantinga concludes] is that there is no reason at all for accepting (it): it is no more than a bit of intellectual imperialism on the part of the foundationalist. He means to commit himself to reason and to nothing more; he therefore declares irrational any noetic structure that contains more—belief in God for example—in its foundations. But here there is no reason for the theist to follow his example; the believer is not obliged to take his word for it.

A mature theist commits himself to belief in God: this means that he accepts belief in God as basic. There is nothing, Plantinga affirms, contrary to reason or irrational in so doing. Let us consider how far Plantinga has succeeded in establishing this point.

I agree with Plantinga that the phenomenological feature which some philosophers have seen as a mark of self-evidence is not a guarantee of truth: a proposition I entertain may possess an evident luster and be false for all that. Again, however vehement an impulse I may have to assent to a sentence I mentally rehearse, the sentence may be false or even sheer nonsense. There is no threshold of vividness or critical degree of compulsiveness which ensures truth. We can record this philosophical insight either by saying that self-evidence is no guarantee of truth, or by saying that not everything that appears self-evident is genuinely self-evident. I share Plantinga's preference for the second alternative, reserving 'self-evident' for propositions which are both vividly assent-compelling and true. The evident lustre is the appearance of self-evidence: only if it attaches to true propositions is it really self-evidence.

I agree also with Plantinga's principal point that it may be rational to accept a proposition though it is neither self-evident nor evident to the senses, nor held on the basis of any reasons. There are many such propositions that I hold myself: such as, that I am awake, that human beings sleep and die, that there is a continent called Australia where I have never been; that there have been Christians for about two thousand years. I claim that I am rational in accepting all these propositions, and in no way guilty of credulity.

My complaint to Plantinga is that one must go much further that he has done if one is to make any substantial contribution to answering the question 'Is belief in God rational?' 'So far' he says at the end of his paper 'we have found no reason at all for excluding belief in God from the foundations. So far we have found no reason at all for believing that belief in God cannot be basic in a rational noetic structure.' I agree that belief in God is not shown to be irrational merely because it is a belief which is not based on reasons while not having as its object a proposition which is self-evident or evident to the senses. But Plantinga has not shown us why what goes for belief in the proposition 'there is a God' may not go for belief in any proposition whatever.[3] For all he has shown there would be nothing irrational in a noetic structure which included among its foundations 'there is no God'. W. K. Clifford could include among *his* foundations the proposition that a noetic structure including belief in God as basic is irrational; and he could do so with perfect propriety provided that he did not go on, as in fact he did, to subscribe to a theory which said that only self-evident and incorrigible propositions could properly be included in the foundations.

We may wonder whether there may not be criticisms of the rationality of people's beliefs based on grounds other than the complaint that their basic beliefs go beyond the austere limits of what is self-evident and manifest to the senses. If not, then it seems we must accept with a shrug that different people's noetic structures may differ in just the way that one man's meat is another man's poison. A's noetic foundations include 'God exists', B's include 'there is no God', C's include 'I am Napoleon', and D's include a design for perpetual motion. Should we worry about this? Do we not glory in being a tolerant and pluralistic society?

It may be that the human condition is as bad as this: but surely we should try harder to see whether we can give a better account of rationality. Just because the criterion for correct belief given by the classical foundationalist fails, we should not conclude prematurely that no criterion can be given which will help to distinguish between rational and irrational beliefs, between sense and folly, between sanity and madness. In the next lecture I shall try to offer such a criterion; in the remainder of this I shall offer some prolegomena to guide us in the search.

Let us go back to the propositions which I claimed to believe rationally for no reason. You may have been surprised by some of the examples I gave. No doubt the propositions were not self-evident and did not report anything that was manifest to the senses: but surely I can give reasons for my beliefs that human beings die, that there is an Australia, that there have been Christians for roughly two millennia. And if I cannot offer reasons to convince you that I am awake, it is only because if you do not already believe that, you will not take anything coming out of my mouth as constituting the offering of reasons.

This last point brings out an important feature of the activity of giving reasons. If I am to give someone a reason for believing that *p*, it is not enough that I should point to a proposition *q* which he accepts and which entails *p* (as 'I am giving you reasons' entails 'I am awake'). Something further is necessary, which is difficult to give a precise philosophical account of, but which Aristotle summed up when he said that the premises of an informative piece of reasoning had to be 'better known than' the conclusion.

So too in my own case: *p* can only be the reason, or a reason, why I hold *q* if *p* is in a more basic position in my noetic structure than *q*. It is because of this that I maintain that there are no reasons on the basis of which I believe such propositions as that 'there is an Australia'.

If I were asked to think of reasons why I believe in Australia, no doubt there might come to mind such things as that I have often seen the continent marked on maps, I have friends who have lived there, I have had letters from there, seen planes depart thither, seen pictures of Australian cities and deserts, drunk Australian wine, seen Australian animals in zoos, and so on.

Two things are striking about this list. The first is that each item, taken in turn, provides on the whole rather slender evidence for the existence of Australia. All I saw, for instance, was the word 'Australia' on the label of the wine bottle, or on the kangaroo's cage. However, it might be urged, taken collectively, these rather flimsy strands of evidence combine to provide substantial support for the proposition.

But the second, more important feature to notice is this. If any one of the 'reasons' for believing in Australia turned out to be false, even if *all* the considerations I could mention proved illusory, much less of my noetic structure would collapse than if it turned out that Australia did not exist. I have only hazy and fragmentary memories of most of the things I might cite as reasons, and most of the things which have, over my life, led to my present firm conviction of the existence of Australia have long been forgotten. Even those which I remember clearly—like looking at an atlas last night—are far less fundamental to my noetic structure than the proposition that there is an Australia. If the atlas had not contained a map of Australia, I would have regarded this as a defect in the atlas, not as disconfirming my belief in the continent; conversely, if the atlas had shown a large continent in the Pacific Ocean at the latitude of California, I would not have taken this as evidence for the existence of such a continent, but rather as indicating that the atlas was not what it seemed. No doubt I cannot seriously contemplate the possibility that all the atlases I have seen have been spoofs cruelly placed in my way, or that I have quite misunderstood the nature of atlases. No doubt so much of my noetic structure would collapse if I came to suspect this that I cannot conceive of coming to regard all the evidence in atlases for the existence of Australia as dubious. But however that may be, the collapse would be minor compared with the havoc in my noetic structure if—*per impossibile*—it turned out that Australia did not exist.

That is why I say that my belief in the existence of Australia is not based on reasons. There are no other beliefs which I have which could be used to support the claim that Australia exists which are better known to me, more firmly established in my noetic structure, than is that proposition itself. If there was any conflict between the types of information which I could give to support my belief in Australia—travellers' tales, or works of geographical reference or the like—it would be these latter, and not my belief in Australia which would have to give way. The proposition that there is an Australia sets the standard by which anything that might be offered as evidence on the topic would be judged credible or incredible. This shows that the proposition is not, in my noetic structure, related to such items in the way that conclusions are related to evidence.

I do not claim that no one's belief in the existence of Australia could be based on reasons, or that my own was never at any time of my life based on reasons. The belief in Australia in a young child today, or of an educated European adult in the eighteenth century, rests on reasons. The basic role of an item in a noetic structure is something which is relative to persons and to times. This is an important point to bear in mind in the context of belief in God. Suppose that Plantinga is justified in saying that for him belief in God is basic: that it is something which he quite properly believes without any reasons. Then his belief in God is not based on any reasons. But that would not prevent him from being able to give reasons to me why I should believe in God. For belief in God is not basic for me as belief in Australia is; and Plantinga might be able to show me that the existence of God was entailed by propositions which are basic for me—which are, perhaps, evident to the senses, like 'some things move' and the other premises of Aquinas' five ways.

Though a belief may be basic for one person and not for another, there are some beliefs which must be basic for everyone. Among my basic beliefs is the belief that other human beings sleep. If this is false, then my whole noetic structure collapses; this is something I know if I know anything at all. This, which I can say of myself, all other sane human beings can say also of themselves. If a belief of this kind were to turn out mistaken, one's entire noetic structure, including the whole methodology of distinguishing true from false, would completely collapse. If any beliefs deserve to be called foundations of knowledge, these surely do.

Let me try to suppose that no one else has ever slept: that throughout my life anyone who has appeared to me to be sleeping has in fact been awake, and that everyone has been united against me in a gigantic and unanimous hoax. If I could seriously entertain that supposition, what reason would I have to trust anything I have ever been told by others, or to trust the ways I was taught to tell one thing from another, or the meanings I have been told of the words I use? To be sure, my identification of objects and my verbal usage appear to have been corroborated over the years by the constant agreement of others; but if they are all leagued against me with the degree of skill and resolution that this supposition implies, then this reinforcement too could perhaps be the result of malevolent and unremitting stage management. Of course this whole train of supposition is literally insane: anyone who pursued it seriously would be quite mad.

Because of this, my belief in a fundamental truth such as this is unshakeable. There can never be any reason for disbelieving it, since any candidate for being such a reason would be something which called in question the possibility of there being any such thing as evidence at all. How am I supposed to acquire evidence for the universal hoax? Someone tells me 'Whenever you have seen anyone asleep, you have been deceived: we have all been fooling you all along.' Would that not be evidence that *he* was mad, if he seriously persisted in the suggestion? Suppose everyone tells me the same story. Well, if I begin to think that is what they are doing, I shall have to give up the idea that I understand what kind of thing human beings are or do or say.

Fundamental truths such as the propositions that human beings sleep and die form a class of propositions which it is clearly rational to believe without evidence, in addition to the classes admitted by classical foundationalism. Such truths are neither self-evident nor evident to the senses: yet they are not believed on the basis of evidence, nor is belief in them a mark of credulity. One cannot give evidence for them, either to oneself or to others: to point to individual slumbers and deaths can only illustrate, and cannot provide evidence for, the general truths. For anyone who is old enough to follow an inductive argument already knows that human beings sleep and die, and will use that knowledge as a standard for testing any individual's claim to be exempt from the need for sleep or the necessity of death. In the noetic structure of anyone who has reached the use of reason such truths have a role which is incompatible with their resting as conclusions on the basis of evidence which is better known.

We have identified, therefore, a further condition which must be added to the classical foundationalist's conditions for rationality if we are to hope to characterize that virtue. It is not, however, an addition which will enable us to say without further ado that belief in God is rational, because the proposition that God exists is not one which must be basic to the belief-structure of every rational human being. Even if it is something which *can* be properly held as basic, it is not, like the fundamental truths we have just been considering, something which *must* be so held. We must consider further the criteria for assessing the rationality of belief. In my next lecture I shall attempt to formulate a definition of the virtue of rationality: a definition which will not have the attractive simplicity of classical foundationalism but which will, hopefully, be free of the self-stultifying quality of that definition.

Like Plantinga, we have rejected the classical definition of rationality as the proportioning of one's belief to the evidence. We have, however, refused to give up at that point the search for the *orthos logos* of the virtue. We have agreed with Plantinga that it is not only self-evident or sensibly manifest propositions which can be properly believed without reasons; but we have not abandoned hope of finding a serviceable characterization of the class of propositions which are thus properly basic. In the next lecture I hope to offer such a characterization, and thus to offer a definition of rationality which will provide a framework for the justification of belief.

# 2 The Justification of Beliefs

We have begun to consider the conditions in which it was rational to hold particular beliefs. We rejected the claim of classical foundationalism according to which the beliefs of a rational person should consist only of propositions which are evident to the senses, self-evident, or else derived from such propositions by a process of reasoning. Such a theory appears to be self-refuting, in that this criterion for rational belief seems to be itself neither self-evident nor evident to the senses, nor is it easy to see by what process of reasoning it could be derived from such premises. Moreover, it appeared possible to furnish incontrovertible examples of propositions which are rationally believed without evidence, while being neither self-evident nor evident to the senses. Some of these, such as the proposition that human beings sleep, are believed without evidence by everybody who believes them; the difficulty of providing evidence for them arises not from their obscurity but their obviousness; there is nothing more certain which could be offered in support of them. Other propositions, which I instanced in my own case by the proposition that there is an Australia, may be believed by some people on the basis of evidence or testimony, but can rationally be believed by others without evidence, because of the fundamental role which they can play in an individual's noetic structure.

It is not enough merely to show that the foundationalist's canon of rationality is self-refuting, or to give examples of incontrovertibly rational beliefs which fail to satisfy his criterion. For this will not be of any assistance in assessing the rationality of beliefs whose status is a matter of controversy, such as belief in the existence of God. That there are some beliefs which are properly held without evidence does not show that the proposition that God exists can be rationally assented to without evidence. In today's lecture I will try to take a step toward assessing the rationality of belief in God by attempting to formulate a criterion for the rational acceptance of belief in general: in later lectures I will apply the criterion so formulated to the particular cases of belief in God and faith in divine revelation.

A criterion for rationality is most helpfully formulated in two stages. First of all, we need to construct a canon for the rational acceptance of a belief as basic, that is to say, for the belief in a proposition without evidence. Second, we need to articulate the ways in which nonbasic beliefs can be based on the evidence of the basic propositions, or otherwise derived from them. In the present lecture I will try to do these things in turn, though I will spend much the greater part of the time on doing the first.

First, then, the criterion for basic beliefs. I offer the following, starting from and expanding the inadequate criterion offered by the foundationalist. A belief is properly basic, I claim, if and only if it is

—Self-evident or fundamental
—Evident to the senses or to memory
—Defensible by argument, inquiry, or performance.

This is a complicated criterion: its value, if any, will obviously depend on how the individual clauses of the definition are spelled out in the sequel. It is a criterion which has none of the attractive simplicity of the foundationalist's canon of rationality. On the other hand it has one advantage over the foundationalist's canon: it is not obviously self-refuting as that one was. While this criterion is neither self-evident nor evident to the senses, it is not necessarily impossible to defend it by argument and inquiry. Indeed I hope to go on and show you that it is so defensible. But of course I do not hold it, nor will I invite you to accept it, as a basic belief: I endeavour to persuade you to accept it only after reflecting on the philosophical considerations I shall lay before you. While not self-refuting like strong foundationalism, my proposed criterion, on the other hand, is not as hospitable to lunacy as an open-door policy of laying down no general conditions or restrictions on what can count as basic.

## 1. Propositions Which Are Self-Evident or Fundamental

To the foundationalist's category of the self-evident our criterion adds the class of fundamental beliefs, the beliefs which are basic in the noetic structure of every rational human being. Beliefs of this kind were subjected to illuminating scrutiny in Wittgenstein's posthumous work *On Certainty*, the treatise on which he was working at the time of his death. Wittgenstein was concerned with a particular class of proposition. These were propositions which, while concerning material objects such as the moon, or human beings, were yet not empirical propositions in the sense of being believed on the evidence of sense-perception. They were propositions which were not self-evident in the way that the a priori propositions of logic and mathematics claim to be, and yet are not believed on the basis of other propositions because there are no other more evident propositions on which they could be based. As examples of such propositions Wittgenstein gives 'The earth has existed for many years past' 'cats do not grow on trees' 'human beings have forebears'. It was as an example of this kind of proposition that I offered earlier the proposition that human beings sleep.

I called these propositions 'fundamental' because they are universally held as basic: all those who hold them do so not on the basis of other propositions. I also said: 'If any propositions deserve to be called the foundations of knowledge, these do.' But there is more to being a foundation than merely being basic: a proposition is a foundation if it is not only unbased in itself but also serves as a basis for other propositions. And it is not at all the case that all propositions which are basic serve as a basis for others. The belief that there is a fly walking up the window pane is basic for me as I write this—it is manifest to my senses. But it does not serve as a foundation for any significant part of my noetic structure, and if it turned out false, little damage would be done to the web of my beliefs. Propositions which are universally basic are fundamental in a different way, in that they could not be given up without causing havoc in our noetic structures; but there is something misleading in regarding them as foundations of knowledge.

Wittgenstein wrote:

> I want to say: propositions of the form of empirical propositions, and not only propositions of logic, form the foundations of all operating with thoughts (with language) . . .

He at once went on to correct himself:

> In this remark the expression 'propositions of the form of empirical propositions' is itself thoroughly bad; the propositions in question are statements about material objects. And they do not serve as foundations in the same way as hypotheses which, if they turn out to be false, are replaced by others. (*On Certainty*, pp. 401–2)

A proposition such as 'the earth has existed for many years' is not a foundation in the sense of being a truth from which other truths are deduced, as one might deduce consequences from a hypothesis or an axiom; rather, 'in the entire system of our language-games it belongs to the foundations'. The language-game is not a set of truths or an axiomatic system; it is the whole linguistic institution within which one distinguishes between truth and falsehood. My picture of the world, Wittgenstein says, is 'the inherited background against which I distinguish true and false'. If I regard some proposition as being at the rock bottom of my conviction, 'one might almost say that these foundation walls are carried by the whole house' (*On Certainty*, pp. 94, 246–48). This means that the metaphor of foundations is not wholly apt. Just as the relation of premise to conclusion is inadequate to explain the kind of support that propositions in my noetic structure give to each other, so too is the relation of foundation to building. A building may collapse if the keystone is removed, not just if the foundations sag. The fundamental propositions in my noetic structure stand fast, as the keystone does, because they are held fast by the propositions which lie around them.

The metaphor of the foundations of knowledge has a great hold on the philosophical imagination. It was given that hold by

Descartes: but Descartes has a different metaphor to describe the conduct of epistemology. He speaks of knowledge as a tree—a tree on which can grow wild and luxuriant branches of belief. Cartesian critical doubt can be regarded as a pruning operation on a noetic tree to turn it from a diseased tree into a healthy, fruit-bearing tree of knowledge. No doubt I can prune the higher more wayward branches by sitting on the lower branches; but I must be careful not to cut off the branch on which I am sitting. The fundamental beliefs which, in the other metaphor, are foundation stones, are, in this metaphor, beliefs which can be pruned only in this way: beliefs which can be called into question only by something which calls itself into question. Just so, I suggested earlier that any evidence purporting to show that human beings never slept would have to be something which called into question, *inter alia*, its own evidential value, by casting doubt on the efficacy of human testimony in general or, more likely, on my own state of sanity. The role of the fundamental beliefs is, I consider, better represented by the arboreal metaphor than by the architectural one.

Wittgenstein himself came to prefer a third metaphor: that of a riverbed. This was more appropriate, he thought, because it took account of the way the role of a proposition in a noetic structure may alter with time.

> It might be imagined that some propositions, of the form of empirical propositions, were hardened and functioned as channels for such empirical propositions as were not hardened but fluid; and that this altered with time, in that fluid propositions hardened and hard ones became fluid. . . . The same propositions get treated at one time as something to test by experience and at another as a rule of testing. (*On Certainty*, pp. 96–8)

## 2. Propositions Evident to the Senses or to Memory

To the foundationalist's category of those propositions which are evident to the senses I add those which are evident to memory. I believe that I had a bath in my hotel room this morning; no one else saw me and I did my best to leave no evidence; but I believe firmly, and not on any evidence, that I had a bath. Also I believe that I lived in Rome from the age of 18 to 25 because I remember doing so. This has no doubt left quite a lot of evidence in letters, passport stamps, photographs, other people's memories, and the like, but it is not on the basis of that evidence that I believe it; for me it is something basic.

Classical foundationalists are unlikely to have any objection to the addition of the deliverances of memory to the category of basic beliefs. Some foundationalists, such as Aquinas, actually believed that memory was a sense, an inner sense, so that they would no doubt regard the data of memory as included under the rubric of things manifest to the senses. It is, I believe, a mistake to regard the memory as a sense, as if it were something like a telescope through which one looks at the past. No doubt, however, it is correct to regard the visual, tactile, olfactory, auditory and gustatory memory as being a sensory capacity; and perhaps this is all Aquinas meant. The point is unimportant here, so long as it is recognized that the memory, as well as the five senses, is a source of properly basic propositions.

Some other foundationalists, in more recent times, have thought that truths about the past were inferred, or deduced from memory images or other mnemic data. They would thus not be basic beliefs, but belong in the class of beliefs justified by their relation to basic data. Again, I think this is mistaken; memory is not any kind of inference from something which is more immediately known. When I believe things that I remember I do not carry out any inference from something which is a quality of my present thoughts or images. To do so would be impossible: what feature of my present thoughts could even show that they were about the past, let alone that they were a true record of it? My belief in the past truths I remember is just as basic as my belief in what I now see. Of course I can be mistaken when I think I remember something; equally I can be mistaken when I think I see something. To be basic is not to be incorrigible, and to believe something as basic is not to claim to be infallible. We can all make mistakes; but not every mistake is a mistake in deduction, as the classical foundationalist of this kind claims.

## 3. Propositions Defensible by Argument, Inquiry, and Performance

This third class is the most significant addition I wish to make to the foundationalist's categories of rationally basic beliefs. It is, in its own right, the widest class of such beliefs; and it is—or so I shall go on to claim—the one which is most relevant to the consideration of the rationality of theism.

We must first make a distinction between the reasons via which a belief is acquired, the reasons on which it is held, and the reasons by which it may be defended. These three sets of reasons may be quite different, without any irrationality, incoherence, or hypocrisy. I may have acquired the belief that X's marriage is breaking up by listening to lunchtime gossip; I may now hold that belief on the basis of a late night heart-to-heart conversation with X's spouse; I may defend the belief to incredulous friends of the once happy couple by pointing to the list of divorce petitions currently filed. In the case of many of our beliefs we have very likely forgotten how we acquired them, and on the basis of what reasons, if any, we came to believe them. That does not mean that we do not hold them now on the basis of reasons.

I believe that the American Revolution was, on balance, a good thing. I cannot remember when I first began to believe this, or indeed whether I have believed it ever since I first heard of the American Revolution. (I suspect not; I seem to remember my elementary school textbook taking a poor view of the Boston Tea Party.) *A fortiori* I cannot remember for what reasons I came to believe it; but there are lots of reasons why I now believe it. And these, unlike the 'reasons' I mentioned in my last lecture for believing in Australia, are reasons for myself and not just reasons I could offer to other people, such as my children, to convince them of the merits of the American Revolution. If the reasons were called in question by a skilled advocate of colonial rule, I might be willing to change my mind on the topic. In the case of my belief in Australia, on the other hand, I no longer have reasons, though no doubt I acquired it on the basis of reasons. None the less I can defend the belief to others by offering considerations which, while not providing reasons for me because they are not better known to me than the conclusion is, might reasonably provide reasons for other people whose noetic structure did not afford the existence of Australia such an assured place.

So a belief may be basic in the sense of not being held on the basis of reason, but yet defensible to others by the giving of reasons.

In my candidate criterion for rationality I have distinguished between three different ways in which a belief may be thus defensible. It may be defensible by argument, or it may be defensible by inquiry, or it may be defensible by performance.

It will be clear enough what is meant by saying that a belief may be defensible by argument. A can defend his belief in *p* to B if B questions it, by offering B propositions *q.r.s.* which B accepts and which either entail or make probable the truth of *p*. But why do I distinguish between argument and inquiry? Pursuing a line of argument may be, perhaps, the pursuit of a form of inquiry. The reason for the distinction is this. If A's belief is challenged by B, A may already be in possession of sufficient information to be able to show B the truth of *p*; all A has to do is to produce the information from his own stock, and draw to B's attention that it provides convincing evidence for *p*. But it may be that A does not have this information at hand, and has to take steps to acquire or recover it. In this case his defence of his belief must involve inquiry as well as argument.

The former is likely to be the case where the belief is one held as basic because it is better known than any available supporting information. The latter is likely to be the case with the more trivial basic beliefs which we acquire every moment of our lives while quickly forgetting how we came by them. But basic beliefs of various levels of embedding may be defensible by argument and inquiry, and if so defensible are perfectly rational.

If a belief does not belong to the other two categories of properly basic belief then, I claim, it is not rational to hold it unless it can be defended in the appropriate way. But to whom is it to be defended, and how successful must the defence be? Must the believer be able to defend the belief to just everybody, and succeed in convincing each challenger of its truth? That would be an excessively severe requirement for rationality. On the other hand, the mere fact that somebody can defend a belief does not make it rational for just anybody to hold it. If wise and well-informed A can defend the belief that *p*, that does not *eo ipso* make it rational for any stupid ignoramus B to subscribe to *p* even though he has never heard of A and has no idea how A would justify the belief.

For a belief to be properly basic for a believer B, B must himself be capable of defending it against those who are likely to challenge it. The believer's defence may take the form of referring the challenger to someone else who can conduct the argument or undertake the inquiry. But even in this case the believer must know who, or what kind of person, has the appropriate competence. Otherwise the relation between B's belief and the third-party argument and inquiry is too remote and coincidental to provide a rational defence of his own basic belief.

We may distinguish between primary and secondary inquiry. There is a difference between working out the square root of 3.4567 and looking it up in a table; there is a difference between looking out of the window to see whether it is snowing, and turning on the radio while lying in bed to find out whether it is snowing. Equally there is a difference between carrying out an autopsy oneself, and reading the pathologist's report; between doing experiments in physics, and reading an up-to-date textbook on the subject. The first line of inquiry is what I call primary inquiry; the second in each pair is what I call secondary inquiry: inquiry which is parasitic on other people's inquiries.

The argument or inquiry which defends a basic belief need not be so successful as to convince the challenger: he may be an unduly difficult person to convince, or perhaps the matter is one on which there is no knockdown argument. But the defence must, by argument or primary or secondary inquiry, produce evidence for the belief appropriate to the believer's degree of commitment. It is in this way, at last, that we make a link between rational belief and evidence, and do justice to the grain of truth in the foundationalist's claim that belief must be in proportion to evidence.

Besides argument and inquiry I have listed a third way in which a basic belief may be defensible: namely, by performance. There are certain fields where the judgement of a person with a special skill or special experience is rightly accepted as authoritative, even though the judgement is not, and perhaps cannot be, backed up with argument or inquiry. I am told that some people in various parts of the world have remarkable skill as water-diviners, or dowsers, though neither they nor anyone else can give an account which will stand up to scrutiny of the methods by which they claim to detect water underground. If such stories are true, then it is reasonable to believe in the presence of water on the say-so of a diviner. This belief is, of course, not basic; the bystander believes on the evidence of the diviner's past record. But the diviner's own belief is not based on an inductive argument of that kind, but is basic. It is none the less rational even though not defensible by argument or inquiry: it is defensible by performance.

Similarly, old Farmer Giles may be able to feel it in his bones whenever it is going to rain. Whether his feeling is a reasonable belief is to be judged not by whatever he may say by way of rationalizing it, but by his success in actually predicting the weather. A shrewd judge of character may be able to predict that so-and-so will go far, or such a one come to a bad end, without being able to say in support anything which really goes beyond 'You mark my words!' If he really is a shrewd judge, the success of his predictions will justify his and our belief in what he says.

With these explanations, therefore, I offer as a criterion for the rationality of basic belief that it should be a belief which is either evident to the senses or to memory, is self-evident or fundamental, or is in neither of these categories but is defensible by argument, by primary or secondary inquiry, or by successful performance.

Having considered the criteria for properly admitting a belief to one's noetic structure as basic, we need to turn, rather more briefly, to the consideration of the nonbasic beliefs in a noetic structure. These are the beliefs which depend on, are based on, the basic beliefs. The criterion of rationality here must specify the relationship in which these nonbasic beliefs must stand to the basic beliefs.

There are two principal ways in which nonbasic beliefs are derived from basic ones: by inference and by testimony. In each case the basic beliefs provide evidence or support for the nonbasic beliefs. But within the broad categories of inference and testimony there are several different kinds of support provided by basic for nonbasic propositions.

When a nonbasic proposition is derived by inference from a basic one the argument may be of a deductive or inductive kind. That is to say, the nonbasic proposition may be derived as the conclusions of a deductive argument of which basic propositions form the premises; or the nonbasic propositions may be hypotheses which are confirmed or made probable by the truths recorded in basic propositions. Deductive and inductive inference furnish an enormous area for philosophical discussion: I shall, however, say no more about them, but will turn instead to the topic of testimony, which is a much less popular topic of philosophical inquiry.

Belief on testimony, like belief arrived at by inference, takes various forms. On the one hand, there is explicit belief in what we are told by others: someone tells me something, and I trust him and believe what he says because he says it. On the other hand, in addition to explicit testimony there is what we might call implicit testimony, in which others manifest their beliefs not by expressing them in words, but by what they do and what they leave undone. One can believe something upon implicit testimony, believing a proposition because others believe it, though they have never put it into words for one. Much of what we believe—all the things that are taken for granted in society or in our group—is believed on implicit testimony in this way.

Perhaps it is not easy to draw a line between what is believed on the basis of widespread implicit testimony and what is simply accepted as basic or fundamental. Wittgenstein has pointed out that many of our most fundamental beliefs about the nature of the universe were not learnt but were rather swallowed down with things which we did explicitly learn. But there is no doubt that testimony plays an important part in the build-up of our world of belief; a more important part than inference whether deductive or inductive.

Professor Anscombe has remarked how much of our world picture is given us by testimony.

Nor is what testimony gives us entirely a detachable part like the thick fringe of fat on a chunk of steak. It is more like the flecks and streaks of fat that are often distributed throughout good meat; though there are lumps of fat as well.

If we say, for instance, that we know there was an American edition of a book because we have seen it, think how much reliance on believing what we have been told lies behind recognizing a publisher's imprint for what it is. My knowledge that I am now in New York, even though I have seen quite a lot of it, is of course based on all kinds of testimony: on maps, notices, on what people have told me. Even those who have lived here all their lives know largely by testimony that this is New York: as they grew up from babyhood that was what they were told by others—the name of the place where they live and its location in the world's geography.

I believe that the conditions laid down here give an adequate account of the rationality of holding any particular belief. Given that someone holds a belief we can ask: Does he hold it as basic? If so, does it fit into one of the three categories of properly basic belief? If not, is it derived by argument or via testimony from some properly basic belief?

However, something else is needed to complete an account of the virtue of rationality. For the conditions considered have all been cognitive, and have not taken account of the volitional side of belief or the relation between belief and value. Moreover a test of the rationality of individual beliefs does not suffice as a test of the rationality of a complete noetic structure. Each one of a person's beliefs might pass the test of rationality as we have outlined it, and yet his system of beliefs as a whole be wildly irrational. These two points are connected together, as I shall explain.

Belief is a cognitive state of mind: it is not an affective state like an emotion or a desire. None the less, it involves an attitude: an attitude towards further evidence on the topic in hand. This is true even more so of knowledge. If I claim to know something—in the strict sense of knowledge which has exercised philosophers through the centuries—I am claiming to be in a position to disregard evidence to the contrary. Of course I may have to consider evidence in favour of something I know to be false in order to help a deluded friend, or to get clear about a philosophical theory, or to avoid offending a benefactor; but I do not have to consider evidence in order to inform myself better on the topic in hand; I am entitled to epistemological disregard. (Of course I may be mistaken in thinking that I am in a position to disregard conflicting evidence; but if so, I am mistaken also in thinking that I know.)

Not all beliefs involve certainty of this kind. Knowledge is not just justified true belief, because a justification which might be adequate for the rational holding of a belief would not necessarily be adequate for the rational holding of something as certain. But any degree of belief involves a degree of commitment. The degree of commitment may vary: a strongly held belief goes with an attitude of disregard, though not complete disregard, of alleged evidence which conflicts with the belief; an opinion held on balance goes with a much more openminded attitude to the conflicting evidence. And when a belief is based on evidence, the degree of commitment should be, as traditionally claimed, in proportion to the evidence (though, as we shall see, that is not the only factor to be taken into account). Basic beliefs too may be held with varying degrees of commitment, and here the measure of the appropriate degree of commitment is not the evidence but the degree of noetic embedding. So criteria of rationality must take into account not only the content, and the basis, of a person's belief but also the degree of commitment.

The second point which our criterion has left untouched is this. The question may arise whether it is rational to have a belief at all on a topic, or to lack a belief on it. It might be that each of a person's beliefs was rational in the sense of being properly basic, and yet his whole set of beliefs was not. This could be so if he had beliefs only on a set of trivial and unimportant matters and lacked beliefs on all kinds of things on which it was important to have beliefs.

For despite what some philosophers have said, there can clearly be duties and obligations to have beliefs on particular topics. This is clearest, of course, in the case of professional people who have a duty to be informed on matters concerned with their profession.

But all human beings need information on thousands of things which are essential for the conduct of daily life, and for their cooperation with others and for dealing with emergencies. It might be said that it is wrong to describe this as an obligation to have beliefs: surely what you need to have is information, namely true belief, rather than just belief. Of course it is, in general, important for us that our beliefs should be true, and this is indeed the reason why rationality is important in belief, not that all rational beliefs are true but that rationality is due process in the pursuit of truth. But just as there are some cases where a decision has to be taken, and it is more important that a decision is taken than that a particular decision should be taken, similarly there are cases where it is more important that someone should have a belief than that it be true. When I sit on a chair I believe that the chair will hold me; that is to say, the thought that it will not does not cross my mind. Better that I should always have this belief—though on occasion it will let me down—than that on each occasion I should first check to make sure the belief is true. That way neurosis lies.

So if we are to assess a person's noetic structure we must not only inquire whether the beliefs he has pass a test of rationality, but whether he has beliefs on topics where it is important that he should have them. The topic of importance is important and difficult: and it is obviously relevant to the discussion of the rationality of theism. Unfortunately it is not possible to separate the importance of a belief from the question of its truth. If there is a God, it seems important that one should believe that there is; but if there is not, is it so important to believe this? As we shall see later, the introduction of considerations of importance complicates the assessment of rationality; but the quest for rationality does not enable us to bypass the question of truth. We shall next consider this in the context of the justification of theism.

# 3 The Defensibility of Theism

In this third lecture I at last attempt to come to grips with the principal problems in the philosophy of religion. The years since the second world war until comparatively recently were a fairly sorry time for philosophy of religion in English speaking countries. It was the period of the rise of linguistic philosophy which was often confused with positivism in the minds of theologians. The rise of this philosophy convinced a number of philosophers that the traditional notion of God was meaningless or self contradictory. Some theologians concluded: 'So much the worse for the concept of God' and attempted to devise a religious atheism; some drew the conclusion 'so much the worse for the notion of self-contradiction' and glorified absurdity with the claim that God was above logic. Neither spectacle was edifying. Those theists who resisted these follies remained on the defensive. Atheist philosophers of religion were more self-confident but not more fertile. A chapter of A. J. Ayer's youthful work *Language, Truth, and Logic* and four pages by Antony Flew on theology and falsification called forth a hundred articles of defensive commentary and tentative refutation. Not since the time of Voltaire have the godly been set on a stir with so little outlay. But the criteria of meaningfulness which Ayer and Flew used to attack theology were, by the end of the seventies, no longer taken seriously, unless by a few theologians. In recent years there has been a revival of interest in philosophy of religion and a return of self-confidence among theistic philosophers.

The time has come to apply our criteria of rationality to the question whether the existence of God can be rationally believed. We will first ask whether it falls into one of our three categories of properly basic belief, and if it does not, we can inquire whether it is something that can be derived in the appropriate manner from beliefs that are properly basic.

Does the existence of God belong in the category of things that are self-evident or fundamental? Distinguished philosophers have thought that the proposition that God exists could be shown to be self-evident. Even the fool who says in his heart that there is no God, St Anselm says, has an idea of God as a being than which no greater can be conceived. But a being than which no greater can be conceived must exist in reality as well as in idea, since to exist in reality is something greater than to exist merely in idea. Descartes also argued that since God was the most perfect being, and existence was a perfection, God must exist. An argument of this kind, which starts from the concept of God, or the meaning of the word 'God', and attempts to show on that basis alone that God exists, is called an ontological argument. Philosophers as distinguished as Aquinas and Kant have endeavoured to show that no ontological argument of this kind can be valid, so that the existence of God cannot be rendered self-evident. When I first came to philosophy some twenty years ago, the ontological argument was regarded as one of the deadest of all philosophical arguments: everyone agreed it was invalid, though there was not similar unanimous agreement as to *why* it was invalid. In the last two decades there have been ingenious and spirited attempts to revive it. I believe that they have failed, and that Aquinas and Kant were right to say that the existence of God is not self-evident.

If it is not self-evident, is the proposition that God exists something that is fundamental, in the sense of being something that is accepted as basic by all those who have an opinion on the matter? There is no doubt that it is possible for an individual to accept the existence of God as basic, and it seems very probable that there have been times and places where the proposition was accepted as basic by whole societies. In the Middle Ages in Europe, for example, the existence of God was probably a basic proposition for the overwhelming majority of believers; this was so even though during the same period it was believed that there existed valid proofs of the existence of God which might be useful in reasoning with unbelievers. But on the other hand it is equally clear that the existence of a God with attributes resembling those of the God of Western theism is not something which has been universally believed by the human race: quite apart from atheists in secular Western societies the adherents of many religions have not been monotheists, and the adherents of some—notably Buddhist—religions are not unequivocally describable as theists at all. Is it even the case that all those who have believed in the existence of the God of Judaism and Christianity and Islam have done so as something basic? I think not. There are many cases in which converts to theism report themselves as having been brought to believe in the existence of God by reasoning and argument. The arguments which converts report as having convinced them may or may not have been sound; but I see no reason to deny their claim that their belief has those reasons as its basis. Others who have been brought up to believe in God, and acquired their belief in the way that basic beliefs are acquired, may well have come at some point to question the belief, to consider the reasons for and against it, and eventually settled down to retain the belief on the basis of that reasoned consideration. It cannot be claimed with any plausibility that all those who believe in the existence of God hold it as a basic belief.

Nor does belief in the existence of God have the kind of unshakeability characteristic of fundamental beliefs: it is not the kind of belief which can only be called into question by something which calls itself into question. No doubt there could be a society in which the existence of God, or of gods, was deeply embedded in everyone's noetic structure in such a way as to be implicit in every kind of inquiry: a society in which the main way in which people tried to find out about future or distant events was by consulting an oracle; in which all trials were carried out by ordeal; in which all cures were carried out by invocation, exorcism, or magic. In such a society, to question the existence of the divine would be to cast doubt on all accepted investigation and diagnosis. But it would only be if all societies were thus that the existence of God could be a fundamental truth in the sense I have defined.

With us, certainly, it is not so. The most earnest believers, in general, do not bring reference to God into their every day or scientific inquiry. Even those who pray to St Anthony to help them find lost keys search their bureau drawers and turn out their pockets in the same way as the rest of us; Chemistry is taught by the same experiments in parochial schools as in nonreligious schools. In the universities God is not mentioned in the scientific textbooks which Christians share with atheists, nor in the project descriptions accompanying requests for research grants. Nor is this something which believers regard as a weakness in their believing in God, a sinful halfheartedness in their religious commitment.

I am not saying that there may not be conflicts between science and religion: there may well be particular issues—e.g. the creation of the world—where the dominant theories in one of the sciences may clash with the dominant interpretation of what is claimed as a religious revelation. What I am claiming is that even where there is an overlap, and a potential clash, between science and revealed religion, this is a clash between two distinct belief systems: the religion is not built into the procedures of scientific inquiry as it would be in the primitive culture that I imagined.

I am aware that even among religious people in our own society who broadly accept the methods and conclusions of contemporary scientific disciplines there are disagreements about the extent to which an appeal to divine intervention is necessary to explain what actually happens in the world. Some Christian men of science operate with a deist model: in creating the world, God sets up once for all a set of natural laws, and the initial conditions upon which these laws operate; all that happens thenceforth (with the possible exception of human choices) is determined by these laws and these initial conditions. Events in history are God's acts only insofar as they are the consequences of this single creative action; if any events are to be specially ascribed to God this can only be because they are miraculous interruptions of the course of nature initially laid down.

There is another view according to which many things are left undetermined by natural laws; not only human actions and decisions, but such things as the course of the weather. Here there is room in history for divine action which is neither the initial determining of the laws which govern our universe, nor yet an intervention in the operation of those laws occurring by way of miracle. But even those who hold this latter view do not regard recourse to God as a method of scientific inquiry: rather they believe that the contingent nature of some events within the purview of science sets limits to the possibility of scientific explanation by such sciences as meteorology, economics, psychology. Belief in the contingency of the weather will not affect the methods of a practising meteorologist: though it may affect the decisions of those who pray for rain or sunshine.

We cannot, then, place belief in the existence of God in the category of fundamental beliefs.

Can we locate the existence of God in the second category of properly basic beliefs—those that are evident to the senses or memory? This reduces to the question whether the existence of God can become evident to the senses. It was necessary, for completeness, to add the category of those that are evident to memory as a category of basic beliefs; but though not everything which is evident to memory is evident to the senses, only what can be evident to the senses can be evident to memory, in the use of 'memory' which is appropriate when we are talking about the use of 'I remember it' to justify a claim to know, or an expression of belief in something.

Memory, we might want to say, is an intellectual as well as a sensory capacity: we can remember a priori truths as well as things we have seen and heard. But to remember something intellectually is simply to have learned it and not yet forgotten it: here, remembering it is just still knowing (believing) it; and to say 'I remember' is simply to reaffirm, not to justify, the claim to know. Whereas, the sensory memory is a source of justification of basic beliefs.

Of course, one can remember something without remembering the sense-experience at the time of what is remembered: I can remember a fact without being able to remember whether I saw it happen, heard it happen, or was merely told about it later. But only what can be sensed can be remembered in the appropriate use of 'remember'. So if someone claims to know that God exists because it is evident to his memory then, unless he subscribes to the Platonic thesis that we can remember things that happened to us in a previous nonbodily experience, he is claiming to have had an experience which made the existence of God 'manifest to the senses'. Is this possible?

If God is an immaterial spirit, and has no body, however ethereal, then God cannot literally be seen with the ordinary senses. Visions of God, such as those attributed to Moses in the Bible, must at best be regarded as the seeing of a miraculous manifestation or symbol of God, not as a literal seeing of an invisible divinity. Even those who take such accounts with the greatest seriousness and reverence usually see them as having the role of a special communication or revelation of himself by God to a favoured servant, rather than as something related to belief in God's existence in the way in which our occasional fleeting sightings of Halley's comet provide evidence for its existence.

The way in which it is popular to claim that the existence of God has the same status as those beliefs which are manifest to the senses is rather this. Some people have, and perhaps all people can have, religious experiences; and religious experiences put us in contact with God. It is true that God, since He has no body, cannot be perceived by the external senses, but we (or at least some of us) have an inner sense which can be trained to focus on God and thus provide irrefutable evidence of His existence.

I think that the expression 'religious experience' is an unfortunate one; not because of anything to do with religion, but because of the confusing nature of the relevant concept of 'experience'. The word is used to cover any item in a person's mental history, whether sense-experience, feeling, emotions, imaginations, dreams and reveries. It thus provides a catchall which includes items of very diverse cognitive status. In particular 'religious experience' includes many different kinds of things, from the most exalted states of those far advanced in mystical pursuits to the sentiments shared by any less than totally hardhearted participant in a religious wedding or funeral.

Sentiments of grief, of guilt, of justification, forgiveness, and exaltation in the context of a religious liturgy or on the occasion of the reading of a sacred text clearly play an important and valued part in the life of a religious believer. The unbeliever may despise them or he may envy them: he can hardly deny their existence or their significance in the lives of those who have them. He may see a distant analogy to them in his own life in the way in which he is moved by reading great literature or stirred by attending or participating in dramatic, musical, and operatic performances. He may find a closer, nonreligious analogy in the sentiments of patriotic citizens at solemn commemorations of national events, or in the emotions of the happily married on attending the wedding of a young couple. On the basis of these analogies he is likely to judge that religious experience in this sense cannot be a basis for belief in God. The sentiments get their significance and profundity from the institutions which provide their context, and not vice-versa. They are not related to these institutions as evidence is to hypothesis. It would be as absurd to argue from the vividness of liturgical exaltation to the existence of God as it would be to argue from the emotion generated by King Lear that he must have existed, to argue from the patriotic fervour one feels at a military parade to the justice of one's country's cause, or from the fond feelings watching newly taken marriage vows to the validity or stability of one's own marriage. We do not argue to truth of Nazism from fervour of Nuremberg rallies; we condemn that fervour because of what we know of Nazism. So in general: we judge the institution to find out what value to put on the fervour it enshrines and unlocks. We do not regard the fervour as justifying the goals and self-descriptions of the institutions. I conclude that religious experience, in the sense of sentiment embedded in religious institutions, cannot make the existence of God manifest to the senses.

What of the rather different kinds of experience claimed by the mystics? Can the experience of mystics be regarded as a perception of God by a secret, interior, sensory capacity? The notion of an 'inner sense' has been popular in the history of philosophy, but it is as confusing as the concept of 'experience'. Many philosophers have regarded the operation of the memory and the imagination as being one kind of inward sense. The implicit comparison with the external senses is misleading.

The external senses—e.g. the five senses of sight, sound, touch, taste, smell—are all discriminatory capacities: e.g. sight is the capacity to tell light from dark, colours from each other; with touch we can tell hot from cold, one shape from another, and so on; with taste we distinguish sweet from sour, cherries from garlic, and so on. For the exercise of each of the senses we have to be in a particularly bodily relationship—differing in a characteristic way from sense to sense—with the object to be sensed: we have to have the cherries in our mouths, turn our head in the right direction to see the star, etc. The senses have organs: that is to say, to each modality of sense-perception there is related a part of the body which can be voluntarily manipulated in ways which affect the operation of the sense: as we turn our heads to listen, feel the shape of something with our hands, open our eyes to see, and so on.

It is these elementary and obvious facts about the senses which mean that their operation is something that can be checked up on and followed by other people. It is these which make sense-perception a public activity that can be shared by others: many of us can look at the same object, hear the same sounds, and so on. These elementary facts are ignored by philosophers who want to talk about imagination and memory as inner senses. To the general incoherence of the notion of an inner sense, there are special difficulties to be added when we consider the idea that God might be perceived by an inner sense.

I have already rejected the idea that God can be perceived by an inner sense of memory independent of the exterior senses. Could we say that the existence of God might be discovered by the exercise of the imagination? The imagination is clearly not a means of acquiring information about the world outside us in the way that the senses are. One cannot discover the way the world is by simply imagining. None the less there is a sense in which we can increase our knowledge of things by using our imagination. It is similar to the way in which we can learn to see things better by drawing them or modelling them. Using our imagination can increase our sensitivity to other people and thus our ability to inform ourselves about what they feel and are likely to do. Works of the imagination may teach us things about human beings; great works of fiction are means by which the human race extends its self-awareness. Could we say that knowledge of God could be acquired by the use of the imagination, in the way our knowledge of ourselves and of our peers grows through storytelling and poetry?

For imagination to be a genuine source of knowledge there has to be some way of distinguishing what is discovered by the imagination from what is created by the imagination. How can we settle whether God is discovered by the imagination or created by it? After all, if there is no God, then God is incalculably the greatest single creation of the human imagination. No other creation of the imagination has been so fertile of ideas, so great an inspiration to philosophy, to literature, to painting, sculpture, architecture, and drama. Set beside the idea of God, the most original inventions of mathematicians and the most unforgettable characters in drama are minor products of the imagination: Hamlet and the square root of minus one pale into insignificance by comparison. But the very fact that an atheist can salute the idea of God as a magnificent work of the human imagination shows that whether God really exists is something which the imagination itself cannot settle. The apparatus of the human mind described by Freud (to take another example) is something whose description does credit to Freud's imaginative genius; but whether it really describes the human mind in a scientific way, or is a newly created mythology, is something which has to be settled outside the realm of the imagination.

Some think that mystics perceive God by a special inner sense, whose object is the divine in the way that sight has light and dark and colour for its object. I do not wish to deny the importance of mystical experience; nor have I any confidence that I can give any adequate account or explanation of it. But I do feel confident that it is misrepresented if it is described as experience of God.

If there is a God with the attributes ascribed to him by Western theism, then he is everlasting, unchanging, and ubiquitous. In relation to such an object there cannot be any activity of discrimination resembling the discriminatory activities of the senses: we cannot have a sixth sense which detects that God is here and not there, as we can see that something is red at one end and not at another, or which detects that God was a moment ago and is not now, as we can hear a noise which suddenly stops. If God is everywhere always, there can be no sense to discriminate the places and times where he is from those where he is not; the whole nature of a sense is an ability to tell differences of this kind. A sense of God would be as absurd as a sense of sight whose only function was to detect a uniform unchanging whiteness or a sense of hearing whose only function was to listen to a single unchanging middle C. Seeing whiteness only makes sense amid telling one colour from another; hearing middle C involves telling it from other notes. One cannot get nearer to or get further from God as one can get nearer to or further from a source of light or sound: one cannot be too early or too late to encounter him as one might be to see or hear something. The whole context within which talk of sense-experience makes sense is lacking in the case of alleged sense-experience of God.

Mystics themselves are as willing to describe mystical experiences in terms of unity of will with God as they are in cognitive terms. But whether the union with God is described in terms of love, or compared with a seeing, or a touching, or a tasting of godhead, it cannot be taken literally as the operation of a sixth sense. For the mode of operation of the alleged faculty differs too much from the mode of operation of genuine senses; and the essential attributes of the alleged object to be sensed differ too greatly from the attributes of any possible object of sensory discrimination.

Those who wish to attribute cognitive value to mystical experiences, whether reached through traditional forms of religious discipline, or allegedly secured by the shortcut of drugs, do better to present mysticism as a nonsensory method of acquiring information than as an extraordinary sense in its own right. But in that case mysticism is seen as a mode of revelation of the divine; and under that description I will consider it in the next lecture. For the present I shall take it as established that the existence of mystical experiences cannot justify the acceptance of the existence of God as basic in the same way as we are justified in accepting as basic those truths which are manifest to the senses.

If the existence of God cannot be accepted as basic in either of the first two categories of properly basic belief, it remains to be seen whether it can be rationally held as basic because it is defensible by argument or inquiry or performance. If this is so, then those who accept the existence of God as something basic may be acting well within their epistemic rights, but only if they can present to others a defence of their belief either by offering arguments or by inquiry, or in some way analogous to justification by performance.

The position that basic belief in the existence of God is defensible in this way has a long history in natural theology. Thomas Aquinas, for instance, though he thought that the existence of God was demonstrable by argument, considered that it was properly held as basic by most believers because of the possibility of offering rational argument in its defense. If this is so, there will be evidence for the existence of God, but it will not be playing the role of evidence in relation to most believers, because for them it will itself be as evident, as well known, as anything which could be offered in evidence for it.

If the existence of God is to be something defensibly held as basic, this will entail that it will be something which can be defended by argument or by inquiry or in some other way. It will be justifiable if any of the traditional arguments for the existence of God can be shown to be sound, and if the traditional arguments against the existence of God can be shown to be invalid.

When I described the forms by which basic beliefs can be shown to be defensible, I listed argument, inquiry, and performance. The category of things defensible by inquiry is not an important one when it is the justification of theism that we have in mind. If there is evidence for the existence of God, it is evidence which is available for everyone. The starting points of traditional proofs of the existence of God have not been recondite facts which only professional men know and which only scientific enterprise could discover. They have been things which everybody knows, as that some things move, or that some things pass out of existence, or that there are relations of cause and effect in the world. But we should look for a moment at the third category of defensible basic beliefs, those which are defensible by performance. A water diviner, I suggested, who has a good record at finding water, is justified in his belief that there is water where the hazel switch twitches in his hand, though neither he nor anyone knows the mechanism, if any, by which he discovers water. I do not, however, think that this provides justification for anyone else to hold as basic the belief that there is water where the dowser says so; on the contrary, anyone else who believes it is doing so—rationally enough—on the basis of an inductive argument from the good track record. Performance provides justification for a basic belief only for the actual performer.

Now who, if anyone, is the relevant performer when the belief in question is belief in the existence of God? One might suggest that the relevant performers are the saints, holy people, those who have led holy lives; perhaps those who prophesy, work miracles, heal the sick. Most saints would regard it, I think, as impious to say that their belief in the existence of God could be justified by pointing to the holiness of their own lives. Even those who work miracles of healing are, after all, *faith* healers; that is to say, the healing is allegedly performed on the basis of the faith. The faith is the precondition of the healing, not something which is justified, to themselves, by the success of the healing. The attitude of mind which would look for such a justification would itself be an impoverishment of the necessary faith. A faith healer may well point to his cures as a reason why others should believe; but if that is the case the belief is being demanded on the basis of the healer's testimony to the content of the belief, and the miracles are being pointed to as the justification of the authority of the witness.

I conclude, then, that belief in the existence of God as a basic belief is something which is justifiable and defensible only if traditional natural theology is a possible discipline, that is to say, if the traditional activity of offering evidence for the existence of God and arguments against disproofs of the existence of God can be successfully carried out.

Interest in the question of the rationality of the belief in the existence of God often originates from a desire to short-circuit the forbidding task of examining the arguments for and against the existence of God. If I am right, there cannot be any such general shortcut. Individuals may believe in the existence of God as something basic, but they are rationally justified in doing so only if it is in general possible to offer sound arguments for the existence of God and to refute arguments against it.

I shall turn in the next and final lecture to whether belief in the existence of God can be rational if it is accepted not as something basic but as something which is derived from something basic. Obviously, if it is accepted on the basis of argument and those arguments are themselves valid and sound then the belief is rational. What is a much more interesting and difficult question is whether belief in God can be valid if based on testimony, and that takes us to the final topic of faith. The world is full of people who offer to bear witness that there is a God and to bear witness that he has planned the salvation, redemption, and judgement of the world. The question of the final lecture is how the rational person should react to that witness, how do we respond rationally without falling into either of the vice of scepticism or that of credulity.

# 4 The Virtue of Faith

In the last lecture I considered belief in the existence of God, and claimed that it could not be rationally accepted as basic on the ground that it was self-evident or fundamental, or that it was manifest to any sense interior or exterior. If it can be rationally believed in as basic, that is because it is defensible by argument or inquiry. If traditional natural theology is a viable enterprise theism is defensible by argument, and thus belief in God is rational, even if held as basic. Whether traditional natural theology *can* be successfully pursued is a large topic to which in these lectures I could not hope to do justice, so that I must content myself with this modest, conditional conclusion.

Can the existence of God be known by testimony? It might be thought that testimony could not be an ultimate source of information: it cannot add to the patrimony of human knowledge, but only circulate existing capital from mind to mind. I do not think this is entirely true: in some areas—the existence of customs and practices, for instance—testimony may be our only source of evidence. But testimony could be given to the existence of God only by reporting experiences of God (and these we have denied to be possible) or arguments to establish God's existence (which will be effective only if natural theology can be successful). The role of testimony in religious belief is not in connection with the existence of God, but in relation to revelation from God. It makes claims, not on belief in God, but on faith in God.

In the theological tradition in which I was brought up faith was contrasted on the one hand with reason and knowledge, and on the other with hope and charity. 'Faith' was used in a narrower sense than 'belief'. Aristotle believed, and perhaps believed with good reason, that there was a divine prime mover unmoved; but his belief was not, according to this tradition, faith in God. So not all belief in God was faith in God. On the other hand, not all faith in God was charity or love of God. Marlowe's Faustus, when he speaks of Christ's blood streaming in the firmament, has long ceased to love God above all things and has no further hope of salvation; yet he retains a belief in the saving value of Christ's death which is faith and not knowledge. He lacks the other theological virtues of hope and charity, but he retains the theological virtue of faith. It is this 'theological virtue' which is the principal topic of this lecture: faith, which is a special kind of belief on theological topics, yet not a belief so special as to find expression in a loving service of God. This is the kind of faith which has often been contrasted with reason. I want to ask now whether faith of this kind is a rational frame of mind, and whether it is, as theologians have claimed, a genuine virtue.

The special nature of the belief that is faith is that it is a belief in something as revealed by God; belief in a proposition on the word of God. Faith, thus defined, is a correlate of revelation; for faith to be possible it must be possible to identify something as the word of God.

In the tradition of which I am speaking the relationship between faith and reason was expounded as follows. It was possible to know, by unaided natural reason, without any appeal to a supernatural revelation, that there is a God. Equally it was possible to know by natural reason certain things about him, as that he was almighty, all-knowing, incapable of deceit, a rewarder of those who lived a good life and a punisher of those who lived badly. Again, it was possible to learn, without appealing to any premise which was not ascertainable and defensible by plain reason, that God has revealed to the human race certain truths about himself which were not ascertainable by reason. The truths which it was alleged had been thus revealed were of various kinds: that Israel was God's chosen people, that there were three persons in one God, that the Eucharist was Christ's body and blood, that the Holy Spirit does not desert the elect, that the wicked would suffer forever in Hell. Those who agreed that there was a divine revelation which called for faith might disagree on the particular content of the revelation.

Equally, there were many different forms in which the revelation might be made, different channels by which it reached the human race from God—perhaps through Moses or the prophets, perhaps through Christ, or the Bible, or the Church, or even in an enlightenment of the individual conscience. What was common to all these different cases was that the believer's faith was belief in certain propositions as having been specially revealed by God. Because the propositions were not demonstrable without appeal to revelation, because they were indeed opaque and might appear repugnant to the unaided human intelligence, faith in these propositions differed from reason. But because the fact of their being revealed could be proved, it was claimed, beyond rational cavil, faith was not in conflict with reason but was a rational state of mind. Faith was a virtue permitting the mind access to truths which would otherwise be beyond its reach.

Faith, so conceived, had a number of important properties. It was intellectual, opaque, rational, free, gratuitous, and certain. It was intellectual in the sense that it was capable of articulation in language: it was not intellectual in the sense of being cold and unemotional (indeed living faith, as contrasted with dead faith, would be alive with love of God and of one's neighbour). But the possibility of articulation in language was a crucial feature of faith: it was indeed the language in which one's state of mind found expression which decided whether one's state of mind was the virtue of faith or the opposed vice of infidelity or heresy, depending on whether it was in accord with creeds and scripture.

Faith was opaque in the sense that at least some of the propositions in which it found expression were incapable of being seen to be true by an intellectual process which did not appeal to the occurrence of a divine revelation. Moreover, at least some of its propositions were so difficult to understand as to be apparently incoherent—they were, in the technical term, 'mysteries' of faith.

Despite being mysterious, faith was rational because publicly ascertainable data existed which provided good reasons for believing that the content of the creeds and scriptures and the authoritative beliefs of the religious community had been revealed by God. The content of revelation was beyond reason's power to reach, but the fact of revelation was something which unaided reason could ascertain.

Though faith was rational faith was also free. That is to say, there was no necessity of nature, nothing in logic or in the makeup of the human mind, which forced any human being to believe, or pevented him from disbelieving. Faith was free, not in the sense that it was optional—it might be required under pain of damnation—but that it was the object of a free choice on the part of the believer.

Faith was gratuitous in the sense that it was a gift of divine grace. That is to say, in the causal explanation of why some people believed and others did not there must enter as an essential element the free decision of God choosing the individuals to believe. This decision was free and one which God could have refrained from making: not just in the sense that so to refrain would be, so to speak, part of his omnipotent repertoire, but also in the sense that no injustice or impropriety would be involved in such a decision.

Finally, faith was certain: it involved a commitment without reserve to the articles of faith; a resolve to disregard evidence conflicting with them. In this, faith was a state of mind resembling knowledge. In the case of faith as in the case of knowledge there might be reasons of a pastoral or academic kind for examining conflicting evidence; but for a true believer there can never be a question of examining conflicting evidence with a view to possible revision of one's commitment to the articles of faith.

So far I have been concerned simply to expound a particular tradition concerning the relation between faith and reason. This tradition I have summed up by saying that according to it faith is intellectual, opaque, rational, free, gratuitous, and certain. Other traditions give different accounts. I have expounded this one not simply because it is the one most familiar to me but also because it is the one which was most explicitly articulated to safeguard the concerns of reason, and therefore the one which it is most appropriate to evaluate when we wish to ask whether faith is a rational frame of mind. Let us now turn from exposition to critical evaluation.

If A believes a certain proposition *p* on the word of God, there are three principal questions which arise. First, what is the relation between A's belief that God exists and his belief that *p*? Second, what is the relation between A's belief that *p* and A's belief that God has revealed that *p*? Third, what is the relation between A's belief that God has revealed that *p* and the historical events that A could point to to justify this belief?

First of all, it seems that faith must presuppose belief in the existence of God. If faith is belief that a proposition is true because God has revealed it, it does not seem possible that one could have faith in the existence of God. When somebody has faith that *p*, his reason for believing that *p* is that God has revealed that *p*. As we have seen in the first lecture, *p* cannot be offered as a reason for *q* unless it is better known than *q*; and *p* cannot be better known than *q* if it is impossible to know *p* without knowing *q*, or impossible to believe *p* without believing *q*. It does not seem possible that someone could believe that God had revealed something without believing that God existed.

I do not deny that one and the same event might provide grounds both for believing that God existed and for believing that he had revealed something. Suppose that the stars were to wheel in their courses and to spell out the sentence: 'The end of the world is at hand.' This would, no doubt, provide reason both for believing that God existed (who else would be powerful enough to control the stars?) and that the end of the world was at hand. I do not say the reason would necessarily be overwhelming. An atheist would, if he thought that was what he saw, no doubt wish to explore other avenues of inquiry before falling on his knees (is it really the stars, or has NASA discovered some new way of producing celestial pyrotechnics?); and even a theist might be in doubt what to think if the message spelt out had been something different, such as 'Drink Coca-Cola' or 'Give ear to the Reverend Moon': is there a powerful demon at work, perhaps, mocking the faithful? Still, such an event would provide a reason for believing in God and believing in the imminent end of the world, even if the reason is not necessarily overwhelming. But in such a case it would be the mode of production of the sentence which would establish the existence of God and the content of the sentence which would reveal the approach of the final days. It would not be a case of believing in the existence of God because God has revealed it, not even if the message spelled out in the skies was 'I am the Lord thy God'.

To say that one could not believe God exists because God has revealed it is not to make any criticism of the traditional account of faith which I expounded. It was no part of that tradition that the believer's belief in the existence of God was itself part of the theological virtue of faith. It was however taught that given the minimal belief in a deity capable of self-revelation, the most fundamental truths about the divinity (e.g. that there is only one God and that He created the world) may in many cases be believed on faith, even though they are naturally ascertainable and can be accepted as a result of rational inquiry.

It follows that it is a precondition of the acceptability of the traditional account of faith and reason that the existence of God is something that can be known or rationally believed. It is not necessary that each believer should have a state of mind which could strictly be called 'knowledge' that God exists. To return to the kind of considerations aired in the first lecture: the little astronomical information I possess could hardly be called 'knowledge' by any strict test; much of it is no doubt hazy misinformation, and even in cases where I am quite confident that I am in the right (as in my belief that Venus is farther away from us than the moon is) I would find it very hard to give a proof which would convince a sceptic. Yet my belief that Venus is farther away from us than the moon is a rational belief because there are other people who know that it is, and I have a good idea where to find them and how I can learn from them.

In the same way many people have believed in God's existence because they were brought up to do so rather than because they had reasons for doing so that would convince an unbeliever. This does not make their belief, and the faith of which it is a precondition, necessarily irrational. It will not do so as long as there are people—

theologians and philosophers, say—who can if the need arises expound to them reasons for believing in God and the existence and role of these experts is known by the believers.

Such is the account given of the condition of 'simple believers' by a theologian such as Thomas Aquinas. This part of his account of the nature of faith does not present difficulty: it accords with the conclusions we reached, from a different viewpoint, about the propriety of holding a belief as basic provided that it was defensible by argument and inquiry. The state of mind of the 'simple believer' is not knowledge, except in the loose sense in which I can talk of my 'knowledge' of astronomy; but it is reasonable belief. Depending on the circumstances of the individual case it may either be basic defensible belief, or nonbasic belief on the testimony of those more expert in natural theology. Always provided, that is, that natural theology is a venture that can be successfully carried out.

I turn to the second question: the relationship between believing $p$ and believing that God has revealed that $p$. One might well ask: if one believes that God has revealed that $p$, how can one fail to believe that $p$? When other people tell us things, we may or may not believe them. Believing someone is not just believing what they say: one might already believe it, or come to see it as true for oneself; it is not just believing something on the strength of their saying so (he may be a habitual liar who always gets things wrong; so when he says that $p$, I believe that $p$, because I know he is both misinformed and intending to deceive). It is to trust someone's word for something. Human beings we may mistrust, but how could anyone in their senses mistrust the word of God, who can neither deceive nor be mistaken?

Aquinas maintained that however certain one might be that God had revealed that $p$, one's belief that $p$ would not amount to knowledge even if one were absolutely correct in believing that God had so revealed. This was because one's opinion about $p$ would be based on an extrinsic criterion rather than on insight into the matter in question. This seems implausible: perhaps Aquinas' opinion was due to excessive deference to a theory of knowledge put forward in the *Posterior Analytics* of Aristotle; or perhaps he was worried that if the believer's state of mind with regard to the mysteries of faith could be regarded as knowledge, faith would no longer be subject to freedom. But this too seems to rest on a misconception of the nature of knowledge: it is all too easy to shut one's mind to what one knows. But it is in fact very unlikely that someone would believe that God had revealed that *p*, and not believe that *p*, though he might try to forget it. What is much more likely is that doubt about *p* will carry with it doubt about whether *p* had really been revealed by God.

The real problem concerning the relation between the belief in an article of faith and the belief that God has revealed it turns on the degree of commitment involved in each belief. Faith, as was mentioned earlier, involves certainty: it is a commitment which is comparable to the commitment which a person has to the propositions which he claims to know. Now is it possible for the belief to have that degree of certainty unless the belief in the fact of revelation has the same certainty? If not, can it be claimed that belief in the fact of a revelation can be rationally held with that degree of certainty? If not, can faith itself be rational?

It is sometimes argued that faith is vicious, because it is a commitment appropriate only to knowledge in the absence of the kind of basis which would justify a claim to knowledge. I have in the past used this argument myself. Reflection on the kinds of considerations spelled out in the earlier lectures has convinced me that the argument will not do. It is wrong to say that one cannot have a commitment to the truth of a proposition as strong as the commitment that goes with knowledge, unless knowledge is present. The fundamental propositions are all held with a degree of commitment as strong as any knowledge claim could carry with it, and yet they are not known in the sense of being supported by justifying reasons. So if faith is vicious, it is not because it is certainty unaccompanied by knowledge.

However, all the cases which convinced me, and I hope convinced you, that there could be justified certainty without knowledge, were cases where the certainty, the unshakeable belief, was not one based on reasons. The reason why the certainty was justified in the absence of the reasons was that the beliefs in question were more certain, were better known than, any reasons on which they might be held. It was precisely because they would have been more certain than any reasons offered that we said they could not be based on reasons; because a reason cannot be less certain than that for which it is offered as a reason.

The arguments we considered earlier showed that it can be rational to believe something with greater firmness than one believes any of the reasons one could offer in support of it, as I believe in the existence of Australia with greater firmness than in any of the reasons I might offer to convince somebody geographically ignorant. But the arguments do not show that one can rationally believe *p* for the reason that *q* with a commitment stronger than one's commitment to *q*. The arguments all concerned cases where it is appropriate to say that *p* is not, or is no longer, believed for reasons.

But this does nothing to solve the difficulty which we have just raised about the degree of commitment involved in faith. For faith is by definition a belief in something for a particular reason: namely, that God has revealed it. Since the alleged divine revelation is the reason for the belief in the article of faith *p*, one cannot rationally believe *p* on faith with a stronger commitment than that of one's belief that God has revealed that *p*. The crucial question, then, is whether this can be believed rationally with the unshakeable commitment faith demands.

Belief in a divine revelation has two elements: first, that there is a God who can reveal himself; second, that certain historical events constitute an actual revelation. The first is something which we have agreed can rationally be believed as basic, if it is defensible, say, by natural theology. But can it be believed not only as basic but as unshakeably certain? This is something which we have not shown any reason to believe. Ought the belief not to be accompanied with a degree of reserve and caution? And if so will it not be inadequate to provide a foothold for faith? Yet we have already agreed that there can be some beliefs held as basic, which it is rational to hold unshakeably, even though they are not fundamental in the sense of being rationally accepted by all who accept them. Might the existence of God be one of these?

Whether this was so or not would depend on the form of the proof of the existence of God which established the defensibility of theism. If it were the kind of proof which showed that the existence of God was, as St Paul thought, something so obvious that only ill will could doubt it, then the answer to our question would be yes. It might be thought that if something were obvious, a philosophical proof would be neither possible or necessary. But this need not be so. Philosophers have doubted whether there are bodies external to the mind, and whether there exist minds other than their own; other philosophers have offered proofs of the existence of the external world and of other minds. But the existence of bodies and of human beings other than myself is so obvious that the only form a 'proof' of their existence could really take would be an exposure of the sophistries in the arguments on the other side, arguments which tend to throw doubt on, or interpret away, the obvious truths that I see other bodies and talk to other people. Arguments for the existence of God cannot be quite on the same footing, since nobody literally sees God or talks with God; but they might have the form of showing that there is something sophistical in any of the arguments which seek to cast doubt on the efficacy of the naive theist's conclusion that there must be a God because the world must have been made by somebody. Whether the natural theologian can succeed in doing this is a large and difficult question; I leave it aside as I am leaving aside the more general question whether the existence of God can be proved in any manner. I turn instead to the second part or element of the belief that a revelation has taken place, that certain historical events are to be interpreted as a communication from God.

The question how events can be interpreted as a communication from God is a difficult one: but I want to consider first the prior difficulty: how is it known that the events in question took place at all? According to the traditional view of the nature of faith, events such as the lives and sayings of Jesus and Moses, and the decisions of authoritative organs of the religious communities, are known by the ordinary methods of historical inquiry.

But if this is so, do not the considerations which enabled us to defend faith against being a conviction appropriate only to knowledge in the absence of knowledge provide a difficulty for the justification of the belief in revelation? For the belief in the fact of revelation has to share the certain commitment of the belief in the article of faith itself; but the belief in the fact of revelation involves belief in propositions based on straightforward historical evidence. So the question is now raised: can the historical facts be known with a certainty which justifies a commitment as strong as that of faith?

This question is partly a philosophical question and partly a historical question. The philosophical question is whether there are *any* facts of history which call for a belief of this degree of commitment. The answer to the philosophical question seems to be clearly yes. I believe that Hitler existed with the same kind of commitment that I believe that Australia exists; the belief is basic in the same way in that it is more evident to me than anything I could offer others by way of evidence in favour of it; and it is unshakeable in the same way in that anything which was offered as evidence against it would call its own evidential value in question. And in addition to basic unshakeable historical truths of this kind, there are others which I would claim to know (as that Cicero was once consul of Rome, and that Charles I was beheaded in London).

Now are there facts interpretable as a divine revelation which are known with this degree of certainty? Whether there are facts in the life of the founders of the great religions which are known in the strict sense is a matter for the historian rather than the philosopher to answer; but I would myself not regard as rash a claim to know that Moses led an exodus of Israelites from Egypt, and that Jesus was crucified, and perhaps that the night before he died he took bread and said 'this is my body'. Can the same be said of all those events in their histories which are pointed to by their followers as the justification for the belief that their characteristic doctrines are divinely revealed? I very much doubt it. No doubt it may be reasonably believed that Moses and Jesus did and said many of the things ascribed to them in the Bible; but can it reasonably be believed with a degree of certainty resembling that of knowledge? Unless the relevant stories can be as certain as the commitment which faith demands of the believer, the commitment is, so far forth as it is faith, irrational; and if the belief is a commitment which is rationally in proportion to the support given by the history, it is, so far forth as it is rational, something less than faith.

Someone may, of course, believe that the Bible is a historically totally reliable account because he believes that it is the inspired word of God. He may believe this not on the grounds of ordinary historical probability, but on the authority of the teaching of a church which proclaims the inspiration and inerrancy of the Bible. But if that is so, then he cannot in turn derive the authority of the Church from the authority of the Bible considered as an inspired document: that would be to argue in a circle. He can rest the belief in the authority of the Church only on the Bible as it is judged by the historian; and that, if what we have just said is correct, is not strong enough to bear the weight of an irrevocable commitment to everything that is contained in it, including whatever passages may be pointed to as the charter of the Church in question.

I conclude, then, my inquiry into the rationality of faith with the conclusion that faith is not, as theologians have claimed, a virtue, but a vice, unless a number of conditions can be fulfilled. One of them is that the existence of God can be rationally justified outside faith. Secondly, whatever are the historical events which are pointed to as constituting the divine revelation must be independently established as historically certain with the degree of commitment which one can have in the pieces of historical knowledge of the kind I have mentioned.

I return finally to the question of the justification of belief in God, belief in the existence of God, as contrasted with faith. I realize that at the end of the last lecture I left this question in an unsatisfactory condition in saying that I thought that belief in the existence of God was justified only if natural theology could be carried out successfully, that is, if the arguments in favour of the existence of God were or could be made sound. I gave no answer to the question whether they can. The reason I gave no answer is that I do not know the answer. I do not myself know of any argument for the existence of God which I find convincing; in all of them I think I can find flaws. Equally I do not know of any argument against the existence of God which is totally convincing; in the arguments I know against the existence of God I can equally find flaws. So that my own position on the existence of God is agnostic.

I say 'agnostic' deliberately, and not 'atheist', because I think that the atheist no less than the theist has to scrutinize his beliefs to see to what extent they pass the test of rationality which I outlined at the beginning. It has been argued by Professor Flew that there is a presumption of atheism: in his book with that title, *The Presumption of Atheism*, he says that unless the proofs of the existence of God are valid then one should be an atheist because the burden of proof is on the theist. But his argument for a presumption of atheism does not succeed. Flew distinguishes between positive and negative atheists; a negative atheist is simply someone who is not a theist. Negative atheism, in his view, differs from agnosticism because agnosticism involves thinking it makes sense to say that God exists, and someone might not be a theist because he thought it did not make sense to say this. Flew then says there is a defensible presumption in favour of negative atheism.

I think it is correct to say that there is a presumption in favour of ignorance over knowledge: that is to say, it takes more effort to show that you do know something than it takes to show that you don't know something. But I do not know why one should call the position of ignorance—the position of not knowing whether there is a God, perhaps not knowing whether it even makes sense to say there is one—negative atheism. One could just as well have divided the field in the following way: a positive theist is someone who positively accepts the existence of God, and a negative theist is simply somebody who is not an atheist. In that case, by a simple redefinition, all Flew's arguments would establish a presumption of theism rather than a presumption of atheism—a presumption of negative theism, of course, but then, after all, it was only a presumption of negative atheism which he claimed to establish himself.

But if we drop Flew's insistence that an agnostic must be someone who positively thinks *God* is a coherent concept, rather than someone who does not know this, then the negative theist and the negative atheist are both better called agnostics. Now is there a presumption of agnosticism? Yes, there is, in the sense that it takes more to prove knowledge than to exhibit ignorance. But this methodological presumption, of course, does not necessarily let the agnostic off the hook. An examination candidate may be able to justify fully the claim not to know the answer to one of the questions set, but that won't get him through the examination.

I must return for the last time to the triad of rationality, scepticism and credulity from which I began. Both the theist and the atheist may well be erring on the side of credulity. The position is not that the atheist is more sceptical, necessarily, than the theist. It may well be the case, and indeed from an agnostic's viewpoint it is the case, that both the theist and the atheist are erring by being credulous. They are both believing something, the one a positive proposition, the other a negative proposition, in the absence of the appropriate justification. On the other hand, from the point of view of either the theist or the atheist, the agnostic is erring by scepticism, that is, he has no view on a topic on which he should have a view. In giving an account of the virtue of rationality, I said earlier, it was not enough to list each of one's beliefs and to see whether each of them passed the test of rational acceptability. Something else had to be considered, namely whether someone had beliefs at all on particular topics. This shows that there is something deceptive about the Aristotelian triad of the person who believes too much, the person who believes too little, and the person who believes the right amount. There is great truth in this, but what is misleading about it is that the criterion for erring on the one side is different from the criterion for erring on the other. You err on the side of credulity if you have a belief that does not fall into one of the categories of justified belief; you err on the side of scepticism, not by the same test, but by lacking a belief on a topic on which you should have a belief. And so the final question is whether it is the agnostic who is erring on the side of scepticism or the theist or atheist who is erring on the side of credulity.

At this point one has to make a distinction between necessary and contingent agnosticism. Necessary agnosticism is the belief which many philosophers such as Kant have had that knowledge whether there is a God or not is in some sense impossible because of the limits of the human mind. There are several philosophical arguments to the effect that agnosticism about the existence of God is something which is built into the human condition rightly understood. I find the arguments for that kind of agnosticism as unconvincing as either the arguments for theism or for atheism.

Contrasted with necessary agnosticism is the contingent agnosticism of a man who says: 'I do not know whether there is a God, but perhaps it can be known; I have no proof that it cannot be known.' Contingent agnosticism of that sort involves not knowing whether other people know, or only think they know, that there is a God. When I, from my agnostic viewpoint, look at my theist and atheist colleagues I do not know whether to envy them or pity them. Should I envy them for having a firm belief on a topic on which it is important to have a firm belief, and on which I myself have none? Should I pity them because of the flimsiness of the arguments and considerations which they use to justify their theism or atheism? From my viewpoint they appear as credulous; from their viewpoint, I appear as sceptical. Which of us is rational, I do not know. Whether this is my own tragedy, or part of the human condition, I do not know.

I have shared with you everything I know on the topic of the relationship between faith and reason. On the really important matter, whether there is a God or not, I have nothing to share with you except my ignorance.

# PART TWO
# Proof, Belief, and Understanding

# 5 Is Natural Theology Possible?

Natural theology, it is sometimes said, is neither natural nor theology. It is not theology, but philosophy: it is the philosophical study of questions concerning the existence and nature of God. It is not natural, but highly artificial: it is a discipline which came into existence only after both philosophy and theology had reached a mature stage of their development.

Some philosophers deny that there can be any such thing as natural theology, because, in their view, all talk of God is an idle use of senseless language. But if that is true, it takes philosophical argument to show it; and that argument will itself be, in a broad sense, a form of natural theology.

What is more interesting is that some theologians also find the notion of natural theology repugnant, and claim that it needs to be transformed into something else—say, philosophical theology. It is often claimed that the very idea of natural theology is no older than the time of St Thomas Aquinas, who first set boundaries between the spheres of faith and reason. Older traditions, we are told, managed without the demarcation between faith and reason, which is full of ambiguities. Does reason mean science or wisdom? Is reason to be restricted to the operations of formal logic, as some analytic philosophers seem to believe, or is it to be given the vast amplitude alloted by philosophers such as Hegel? Is not reason involved also in working out the content of faith? In matters concerning God proof is impossible and if it were possible it would not be enough for faith. Therefore, we must give up the hope of separating out faith and reason.

This is a fairly common line of argument among theologians,

and among theologically influenced philosophers. In fact, in my view, the distinction between reason and faith is reasonably clear. Suppose a philosopher, whether Fregean or Hegelian, offers an argument for a theological conclusion. We can ask whether any of the premises of the argument are claimed to record specific divine revelations or not. Are any premises of the argument put forward because they occur in the Bible, or are there premises claimed to have been revealed to the arguer in a vision? Or, on the contrary, are all the premises put forward as facts of observation or truths of philosophy? If the former, we are dealing with revealed theology; if the latter, with natural theology.

Paradigms of natural theology, such as Aquinas' five ways, or the Cartesian arguments of the Third and Fifth Meditation contain no appeal to a sacred book or special revelation. 'Some things move' 'pepper and ginger are hot', to take the premises of two arguments offered by Aquinas, are facts of experience. Similarly, Descartes claims to experience in himself an idea of God. Both Aquinas and Descartes make use not only of facts of experience, but also of general philosophical principles, such as that every moving thing has a mover, or that there must be as much reality in a cause as in an effect. True or false, these principles are put forward as philosophical theorems, not as sacred truths of a mystic revelation. Anselm's ontological argument does refer to the biblical text about the fool who says in his heart that there is no God; but the role of the quotation is purely ornamental. The force of the ontological argument in no way depends on accepting the Bible as authoritative. Where we have a conclusion to God's existence based on facts of experience and principles of philosophy we have a piece of natural theology, successful or unsuccessful as the case may be.

No doubt Aquinas was the first to lay out clearly the difference between natural and revealed theology. But the basic insight behind the distinction is that there is a difference between the kind of argument which may be used by one religious believer to another religious believer, and the kinds of argument which the believer must use to the unbeliever. And this is something which goes back at least to the sermon of Paul on the Areopagus.

It is true that the conclusions for which natural theologians argue—e.g. that God is omnipotent or omniscient—are to be found stated in Scripture. It may indeed be from Scripture that natural theologians got the very idea of the conclusions they set out to prove by reason. None the less, if the proof they offer does not appeal to Scripture, the conclusion is genuinely one of natural theology.

It is true that philosophical reason is used by theologians in the discussion of the articles of their creed, and in drawing conclusions from sacred texts. This does not have any tendency to blur the distinction between natural and revealed theology. Natural theology purports to be theology for which natural reason is sufficient. This is quite consistent with natural reason being necessary, but insufficient, for reaching conclusions about mysteries of faith. Both kinds of theologians use reason; the natural theologian claims to use reason (and experience) alone. It is the use of revelation which distinguishes the revealed theologian from the natural theologian; it is the use of reason which distinguishes the theologian from the enthusiast.

When we speak of natural theology as the product of 'unaided reason' this is, admittedly, ambiguous. In one sense it may simply be a way of putting what we have already said, that in arguing for its conclusions natural theology does not call in aid any revealed premises. In another sense, it may mean that the natural theologian reaches his conclusions without the aid of divine grace.

We may raise the question whether there is such a thing as divine grace, and if so what is its nature, and how it fits in to the economy of epistemology and salvation. But the question, if raised, seems to be clearly a question belonging to revealed theology and not to natural theology. However, this much seems clear: that when we talk about 'unaided reason' in the first sense, we are talking about the premises from which reason reaches its conclusion, and we are talking about logical relationships. On the other hand, when we contrast unaided reason with the aid of grace we have moved from the realm of reasons to causes: we are talking about the causal, not the logical, antecedents of the reasoning process.

Let us take a crude analogy. A runner may be able to reach a certain speed by the unaided power of his nature; a speed greater than that is beyond his unaided power; but perhaps he can reach it with the aid of anabolic steroids. Similarly, it may be claimed that there are certain beliefs which it is beyond the unaided power of reason to attain to, but which can be reached with the aid of grace. Among such beliefs are the mysteries of faith.

It has been a matter of dispute among theologians whether grace was needed not just for the faith which guides the revealed theologian, but also to enable the natural theologian to reach his conclusions. The Augustinian tradition stressed the necessity for divine illumination at every stage of the mind's journey towards the knowledge of God. What has been a near unanimous view among Christians is that grace was needed for the saving faith in the mysteries of revelation such as the Trinity and the Incarnation.

Faith, in the theological sense, is belief in something on the word of God. Faith is different from the kind of belief in the existence of God which natural theology, if successful, produces. The faithful believer takes God's word for many things; but one cannot take God's word for it that He exists. Belief in God, in this sense, is not part of faith, but is presupposed by it. Some theologians have described it as being one of the 'preambles' of faith.

A contrast is sometimes made between Aquinas' search for proof and the Augustinian notion of faith seeking understanding, illustrated by Augustine's frequent use of a quotation from Isaiah: unless you believe you shall not understand.

On the face of it, the relation between belief and understanding seems to be wrongly stated in the Augustinian tradition. In the case of any individual proposition, understanding seems possible without belief, but not belief without understanding. I cannot believe that *p* if I do not understand what is meant by '*p*'; though I may believe that '*p*' is true without this understanding. On the other hand I may well understand what is meant by '*p*' without believing that *p*; as when I understand the meaning of a proposition I know to be false. Many theological doctrines, whether or not they are true, seem perfectly intelligible to the unbeliever: for instance, the doctrine of the resurrection. It may take faith to believe that

Jesus rose on the third day; but there is no difficulty in understanding what is meant by the doctrine.

Some doctrines, no doubt, are very difficult to understand and some may indeed appear to the unbeliever to be unintelligible or self-contradictory. If the doctrines are genuinely unintelligible, then the attitude to them of the devout can hardly be called genuine belief; one cannot really believe nonsense, no matter how devotedly one tries. But if they are not unintelligible, then while the believer may have the stronger motive for seeking to make plain their intelligibility, there seems no reason why an unbelieving philosopher might not also strive to do so. Philosophers spend much of their time exploring the coherence of suppositions which are at least as prima facie implausible as the doctrines of the Trinity or of transubstantiation.

In setting out the structure of natural theology, St Thomas makes a distinction between the beliefs of the learned and the beliefs of the simple. If the belief of the simple believer is to be justified, there must be arguments for the existence of God. For faith to be reasonable, what is necessary is that belief in God should be reasonable. However, the existence of God need not be something that is actually known, in the ordinary sense, much less be the object of *scientia* in the technical sense which Aquinas uses, following Aristotle. But St Thomas seems to be correct in thinking that the belief of the unlearned believer is reasonable only if arguments for the existence of God are available to the believing community.

Natural theology is sometimes defined as consisting in the attempt to provide proofs or arguments for existence of God. This is a rather narrow definition: natural theology should surely also include other things. It should include the attempt to identify some at least of the attributes of God, and to defend the coherence of these attributes with each other, and the compatibility of the existence of God with known facts about the world.

Natural theology must also be concerned with the meaningfulness, and not just with the truth, of statements about God. Unless they are meaningful such statements cannot be true. Natural theology's concern with the truth of statements about God principally turns on the question why one should believe such statements. And the concern may take two forms: of urging that such statements should be believed without argument, or of providing arguments for believing. This goes not only for the natural theologian but also for the natural atheologian, that is to say, the philosopher who uses philosophical arguments to prove the truth of atheism. In saying that there is such a discipline as natural theology I am not begging the question whether the outcome of the discipline is to establish the existence, or the non-existence, of God.

Allowing that arguments for and against the existence of God is only a part of natural theology, we may none the less ask: What is the point of arguments for the existence of God? Such arguments may be devised in order to convince oneself or others of the existence of God. This would, indeed, seem to be the most obvious and important purpose of such arguments; but it is often alleged that historically the main purposes of the arguments have been something different.

In one tradition their purpose is to transform faith into knowledge, where knowledge is thought of as the most estimable epistemic state, a state superior to mere belief. A long religious tradition regards faith as being as strong as the strongest human commitment to any truth. I believe that this tradition is mistaken: while there are degrees of warrant, no argument of natural theology can confer the highest degree of warrant on belief in the existence of God.

There is another traditional view of natural theology. This claims that, in the absence of evidence, theistic belief is wrong: one can be within one's rights in believing in God only if one possesses propositional evidence for belief. On the earlier view, which we may call the Thomist one, argument is needed to turn a good epistemic state into a better one; on this second view it is needed to turn a bad epistemic state into a tolerable one—the theistic arguments are what provide justification for the belief and make it permissible to accept it.

In recent years, Alvin Plantinga has produced a series of arguments against this view of natural theology. There is no duty,

Plantinga claims, to proportion one's belief to evidence; so this function of natural theology is a needless one. Belief in God need not rest on evidence; it may be basic, and quite properly so.

Plantinga's writings form the most substantial recent philosophical attack on the essential significance of natural theology, and I wish to spend the rest of this paper in setting out my agreement and disagreement with him. I agree with Plantinga that a belief may be justified, and indeed may constitute knowledge, in the absence of evidence; there is such a thing as proper basic belief and it may amount to knowledge. In particular I agree that belief in God may—if certain conditions are fulfilled—be properly basic. No doubt it is usually acquired by testimony, but it remains long after the testimony itself has been forgotten. Which of us here remembers on which occasion, and by whom, we were first told of the omnipotence or omniscience of God? Even if the testimony is remembered, belief in God may properly have a strength which is greater than that of other claims which might be made on the same testimony. The grandmother who told me that God was a loving father was the same grandmother who told me that members of the Labour party were simply Communists in disguise. There is in itself nothing improper in the retention of a belief as a basic belief, even if it was acquired on the basis of testimony. Many of my beliefs are of this kind; for instance, my belief that Great Britain is an island.

But this kind of basic belief is proper only on certain conditions. I have tried to spell out the conditions in my lectures on Faith and Reason in the first part of this book. Roughly speaking, a belief can be properly held as basic, without evidence, only if it is rationally *defensible*. If the existence of God is to be something justifiably held as basic, it must be defensible by argument (since in the case of God other forms of defensibility—e.g. by empirical investigation—do not apply). It will be defensible if any of the traditional arguments for the existence of God can be shown to be sound and if the traditional arguments against the existence of God can be shown to be invalid.

Plantinga in a recent paper considers the suggestion that the function of natural theology is to provide not evidence, but warrant, for belief in the existence of God. What is warrant? Plantinga dismisses three theories of warrant—internalism, coherentism, and reliabilism—and presents his own account. Warrant is a matter of a belief's being produced by faculties that are working properly in an appropriate environment. If a belief has warrant for you, then the greater your inclination to believe it the more warrant it has. A belief is proper if it is formed by properly functioning faculties in an appropriate epistemic environment, where the modules of the design plan of the mind involved in its production are aimed at true belief.

Having thus defined warrant, Plantinga asks: is natural theology needed for belief in God to have warrant? Must belief be based on argument? Given Plantinga's account of warrant, this is equivalent to the following question: when people accept belief in God in the basic way, is it the case that sometimes their faculties are functioning properly, and the modules of the design plan governing the formation of this belief are aimed at truth?

People have feelings such as guilt, shame, and awe. These religious experiences contribute to people's feeling impelled to believe in God. Does this ever happen in case of someone with properly functioning faculties? Not, Plantinga observes, if we are to believe Marx or Freud. For Marx, belief in God is an illusion, produced by malfunctioning faculties. Freud would agree that belief in God is an illusion, but perhaps it is an illusion produced by a properly functioning mental module whose function is something other than the production of true beliefs.

But why, Plantinga asks, should we believe Marx or Freud? Why should we not believe Calvin instead? Calvin says that a sense of deity is inscribed in hearts of all, and belief in God is the product of this faculty carrying out its proper function. How can we settle whether Freud, Marx, or Calvin is right here?

The question 'can basic belief in God properly lack warrant?' is a question which cannot be answered neutrally, because of the relation between warrant and proper function. Your view as to what sort of creature a human being is will determine your views as to what proper function is. From a nontheistic perspective it will be natural to think that arguments of natural theology will be needed for belief in God to have warrant. From a certain Christian viewpoint, basic belief does have warrant and natural theology is not needed.

However, the arguments of natural theology can increase warrant, and increase it significantly, perhaps to the point of turning true belief into knowledge. It is true, Plantinga admits in conclusion, that none of the traditional theistic arguments measures up to the standards applied to them; but no other philosophical arguments for interesting conclusions do so either.

Plantinga's considerations about warrant are of great interest in their own right. They seem to me, however, only marginally relevant to the concerns of natural theology. For as Plantinga says, the question whether basic belief in God has warrant cannot be settled independently of the question whether there is a God or not. Someone who wishes to convince another to believe in God cannot produce this conviction by showing that such a belief would be warranted; for the belief is only warranted if the mind is working properly, and she does not know what it is for the mind to work properly unless she knows whether there is a God or not. The doubting believer in God cannot reassure himself that his belief is warranted; for only if there is a God is his belief warranted, and that is what he was beginning to doubt.

It is not clear from Plantinga's exposition whether belief in God is produced by a module which has other purposes, or whether there is a specific God-belief module. If the latter, we may ask whether everyone has one, and whether it is the same in Calvinists as in Buddhists. We may also ask what other equally specific modules there are: is there, for instance, a module whose function is to produce belief in Santa Claus?

It is not obvious how such questions would be settled. But suppose we settle them in the affirmative: does this help us? No doubt if there is God-belief module it is functioning properly if it produces belief in God; similarly, a properly functioning Santa-Claus-module will produce belief in Santa Claus. But as Plantinga's consideration of Freud made clear, we cannot rule out the possibility of belief-producing modules in the design plan of our minds whose function is something different from truth. So even if we accept that believers are exhibiting their proper function in believing in God, we are not thereby committed to accepting that their beliefs are true.

Of course, if there is a God it would no doubt be cruel of him to give people belief-producing-modules on matters of vital importance which were productive of false beliefs; and indeed God, omnipotent though he may be, could not give people a God-belief-producing module which produced the false belief that God exists. But to get any further with this line of thought, we need to find out whether there is a God or not. And it will not sufficce here to make use of our God-belief-producing module, even if we happen to have one. It will surely be necessary to fall back on something like the traditional arguments for the existence of God.

It seems more likely that belief in God, like belief in Santa Claus, is not the result of some specific module, but rather the outcome of a more general faculty. Both beliefs, perhaps, are brought about by our capacity, indeed our tendency, to believe what we are told. The ability to learn from testimony is certainly one of the most precious of human possessions, and Plantinga is right to stress how much of our knowledge is dependent on testimony.

But, in most matters, the chain of testimony must come to an end. We can only learn from testimony what, at some point, was learnt by means other than testimony. We believe eye-witnesses about what they learnt by seeing, and many chains of testimony lead back to past sense-perception. But testimony can rest on something other than sense-perception: we may believe someone who tells us he has proved some abstract theorem, even though our own intelligence may not be adequate to follow the proof. So proof, no less than sense-perception, may provide the foundation for testimony.

If we are dividing the mind into faculties, then the senses and the reason are two great truth-directed capacities. God is imperceptible by the senses: so that if there are only these two truth-directed capacities belief in God can be produced only by reason. Hence, however long the chain of testimony whereby the child learns of the existence of God, if the belief is to be justified it must rest on proof. The metaphor of chain is perhaps inappropriate: testimony is a web rather than a chain. So once again we reach the conclusion of Aquinas: if the beliefs of the believing community are to be justified, proof must be available somewhere within the community.

Plantinga might question several premises in the foregoing argument. He might take literally Calvin's talk of *sensus divinitatis* and claim that God is indeed perceptible to a sense: a special and inner sense, but a sense no less. However, if God is the kind of being described by traditional theology, He could not possibly be discerned by the kind of discriminating faculties that are what we call senses; and the talk of inner sense, whether in divine or secular context, is usually a symptom of philosophical confusion. More plausibly, Plantinga might deny that there are only these two truth-producing faculties or groups of faculties, namely the senses and reason. Perhaps this is what is really meant by talking of *sensus divinitatis*.

At some points Plantinga seems to talk as if emotions are truth-producing faculties. He appears to argue that because we feel shame at our sins and awe and gratitude at the grandeur and beauty of the world, that gives warrant to belief in God. But this is surely perverse. Emotions can certainly be appropriate or inappropriate; but it is the truth of the matter in hand that makes the emotions one or the other; it is not the appropriateness or inappropriateness of the emotion that confers existence on its object or truth on its propositional expression. Atheists sometimes use similar kinds of argument in the opposite direction; they are equally misguided. The God described by Calvin or Jonathan Edwards, they say, is so unutterably horrible that he cannot possibly exist. But there is no argument from horror to non-existence, any more than there is from awe to existence.

There is indeed one faculty which might reasonably by claimed to be a source of truth independent of sense and reason: namely understanding. (Some philosophers have distinguished reason from understanding, understanding being the capacity for insight, while reason is the capacity for argument.) It is understanding which is what we acquire when we can read what we see; when we perceive something as a sign of something else. Theists sometimes claim to read God out of the world; atheists, on this picture, are life-long dyslexics. Now there is undoubtedly such a faculty as understanding, and it is undoubtedly of maximal importance. The difficult thing is to detect when this faculty is properly functioning (as in the scientist who discerns a pattern where all seems disorder to the untrained) and when it is going astray (as in the superstitious or the person who has a system for beating the bank at Monte Carlo). Essentially the understanding is a faculty for producing hypotheses. These, in their turn, need to be checked; and they are checked by nothing other than the senses and the reason. When what is hypothesized is God, the senses cannot provide the check. So we are once again drawn to affirm the necessity of natural theology.

The introduction of proper function as the criterion of warrant does not at all help the believer to convince the unbeliever or the doubter to convince himself. For however strongly the believer, in his believing moments, feels inclined to believe in the existence of God, this will not warrant him in believing in God in the absence of argument—unless there is a God-belief-producing module whose function is truth-producing. And if the doubter feels inclined to doubt the existence of God, that very fact itself, in Plantinga's terms, reduces the degree of warrant which he has for the belief. So Plantinga's theory of warrant, so far from reassuring the doubter, should make those of us inclined to doubt God's existence even more inclined to do so.

As a doubter who would welcome the resolution of his doubts one way or the other I feel more hope of seeking a resolution from traditional natural theology than from the theory of warrant as proper function.

# 6 The Argument from Design and the Problem of Evil

Forty-five years ago the Oxford theologian Austin Farrer published a rich, but since undeservedly neglected, book on rational theology entitled *Finite and Infinite*. He concluded the book with a section entitled 'Dialectic of Rational Theology' in which he classified different arguments for the existence of God.

Every argument for God's existence must start from the world of finites: it takes some distinction within the finite, and claims to show that the co-existence of the elements distinguished is intelligible only if God exists as the ground of such a co-existence. Arguments for the existence of God, Farrer maintained, can never be formally valid syllogisms because of the presence of analogical terms (such as 'cause' 'existent') in the premises and the conclusion. But each argument is designed to elicit a cosmological intuition, by presenting a distinction of elements within the creature which makes us jump to the apprehension of God as the being in whom this distinction is transcended.

Arguments for the existence of God will differ from each other according to the finite distinction taken as the basis of each; but they can differ also in the form the argument takes. Let the finite distinction be of the elements A and B. Then we may (1) take A for granted, and show the addition of B to it as necessarily the effect of divine action (or vice versa) or (2) take neither for granted but exhibit the combination AB as forming a nature so 'composite' that it must be regarded as derivative from that which is 'simple' in this respect.

Farrer applies his scheme to a number of familiar and unfamiliar arguments of rational theology—from the distinction essence-existence, from the distinction actual-possible, from the distinction between intellect and will, and so on. I wish to consider here just one of the applications he makes of his scheme: to the argument which he calls the argument 'from Formality and Informality (Chaos)'.

The world, Farrer says, is a composition of form and chaos, each form struggling to dominate the irrelevance of an environment which is chaos relatively to its formal requirements. It is this which is the basis of the argument from design.

In the 1A form, we presuppose chaos. If the world were through and through coherent design, that would be its nature and no explanation would be required. The mystery is that design should have got such a hold on material lacking form: this must have been imposed from above by a supreme artificer.

> The great difficulty of this argument is the difficulty of presupposing chaos. Chaos is a chaos of forms; stripped of them it is nothing but the spatio-temporal scheme of the interaction of finite forms . . . It seems then that we must presuppose not naked chaos, but a chaos of low-grade forms in order to raise the question, how (since these do not need the higher forms for their existence) the higher were imposed upon so recalcitrant a medium. Yet this way of stating the question has its own absurdity; for if the lowest forms, by themselves formal, can be taken for granted in their chaotic interaction, what fresh principle or fresh difficulty is raised by the interacting of higher forms with one another and with the lower in the same disorder? (p. 276).

Let us change then from the 1A to the 1B pattern, where we presuppose not chaos, but forms.

> Surely [form] must have suffered violence from some external power in being thus chaotically interrelated or juxtaposed. This power must himself be supposed exempt from such juxtaposition. If the former argument, presupposing chaos, were absurd in its premise, this argument is absurd in its conclusion. For why should a being, himself completely 'formal' i.e. harmonious, smash finite form against itself in chaotic destruction? The conflict between the argument of the proof and its conclusion is such that it is not usually known as a proof of God, but as the 'Problem of Evil'. Why should God cause the forms of human and animal existence to break

against one another, and against inanimate nature, producing the most appalling deprivations and injuries in the physical, sensitive, and spiritual orders?

If we advance from pattern 1 to pattern 2, Farrer claims, we both produce an improved version of the Argument from Design, and we are rid, at a stroke, of the venerable Problem of Evil. The pattern 2, we remember, is the one which takes neither form nor chaos for granted. Farrer states this version of the argument as follows.

> Admitting that the finite, as we know it, is a chaos of forms, we may argue as follows: In so far as there is an element of disorder in the universe, this implies some collocations of substance which cannot be derived from the formal principles of these substances nor from a form of their correlation. Accidental collocation is a mere fact, neither the form nor the expression of any finite operation. It ought to be reduced to a real operation on the part of a being not subject to accidental collocation with other things, nor to the accidental collocation of elements within itself. This non-composite being, then, has placed or created composite being.

Even thus reformulated, Farrer maintains, the argument from Design, like all arguments for God's existence, involves formal fallacy. None the less, it can, he maintains, defy the 'Problem of Evil' attack.

> For granted that existence at our level must be splintered, collocated, and accidentally interrelated, it is not a matter of principle just what miseries arise; 'I could believe in God, were it not for cancer' is an absurd contention; for the nature of accident is to be irrational, nor can it be controlled by measure. It is a practical, not a speculative problem: of cancer research, not of theodicy. 'I believe in God because the world is so bad' is as sound an argument as 'I believe in God because the world is so good'. It could not be so bad if it were not so good, since evil is the disease of the good. (p. 278).

Farrer's style is difficult, his terminology often idiosyncratic, and his theory of the relationship between analogical predication and formal fallacy needs careful examination which it will not receive in this paper. None the less, the passages which I have quoted present a number of metaphysical insights which can be detached from their systematic context, and restated in terms which many of us may find more familiar. I shall try today to restate and defend the link which Farrer enunciates between the problem of evil and the argument from design: for I believe this to be an insight of fundamental importance in natural theology, and in particular in natural theodicy.

I say 'in *natural* theodicy': because there can be various theodical disciplines, depending on which version of the Problem of Evil the theologian wishes to dispel. Farrer is concerned, and I shall be concerned, only with the natural problem of evil, and not any of the versions of the supernatural problem of evil.

Let me explain what I mean by this distinction. Let us assume that—as most of the great philosophers throughout history have believed—the world we live in provides us with reason for believing that it is the work of a powerful and good God. Then there is a problem of accounting for the evil it contains, in so far as that can be thought to be traceable to the maker of the world. This is the natural problem of evil which natural theodicy sets out to dispel.

But if we accept that it is possible to know more about God than natural theology provides, then there may be other, perhaps greater, problems of evil. If we believe that God is not just good, but positively loves his creatures, then the existence of natural evils becomes that much more difficult to account for. If a revelation claims that God not only permits natural evils, but imposes on some of his creatures supernatural evils such as eternal punishment, then the supernatural problem of evil takes a particularly excruciating form. To resolve these problems of evil is a task not for the natural theologian or philosopher of religion, but for the dogmatic theologian, for the professional spokesman for the alleged revelation in question. In this paper I shall restrict myself, as Farrer does in his book, to the natural problem of evil and the province of natural theodicy.

It is a feature which is common to the proof from design and the problem of evil that both are arguments which argue from values to facts. The argument from design can be summarized thus:

There is a great deal of good in the world: therefore there is a God.

The problem of evil, when used as an argument against theism, proceeds as follows:

There is a great deal of evil in the world: therefore there is no God.

Those philosophers who are true believers in the logical importance of the fact-value distinction should have no truck with either the proof from design or with the problem of evil. To be sure, it is usually the derivation of values from facts which the fact-value distinction is cried up to exclude. But if values cannot be derived from facts, then facts cannot be derived from values either. Let V be a value judgement, and F a factual statement. If 'If V then F' is a sound principle, then by contraposition so is 'If not F then not V'; hence if a factual statement can be derived from a value judgement, a value judgement can be derived from a factual statement. The fact-value barrier must be a two-way barrier, or no barrier at all.

It may have caused surprise, however, that I stated that the argument from design involves value judgements at all. Does not the argument from design simply take as its starting point the existence of teleological phenomena in the world? Surely all that teleology involves is a particular pattern of explanation, rather than any reference to good and evil?

Certainly, it was thus that teleological explanation was understood by Descartes, who is commonly awarded the credit, or blame, for cleansing science of teleology. Descartes, it is well known, rejected the explanation of gravity in terms of attraction between bodies, on the grounds that this postulated in inert bodies knowledge of a goal or terminus. But Descartes was wrong in seeing end-directedness, in this sense, as the distinguishing mark of teleological explanation. The essence of teleological explanation is not the fact that the explanation is given *ex post*, or by reference to the *terminus ad quem*. It is rather the part played in the explanation by the notion of purpose: the pursuit of good and the avoidance of evil.

Newtonian inertia and Newtonian gravity provide examples of regularities which are not beneficial for the agents which exhibit them: one is a form of *ex ante* explanation, the other *ex post*. All teleological explanation is in terms of the benefit of agents: but within this there are both *ex ante* regularities (like instinctive avoidance behaviour) and *ex post* regularities (like specific habits of nest-building). Of course there are also teleological explanations of non-regular behaviour, such as human intentional action.

The nature of teleological explanation is often misstated both by its critics and its defenders. Critics allege that to accept a teleological explanation is to accept backwards causation: the production of a cause by its effect. But someone who explains behaviour B of agent A by saying that it is what is required in the circumstances to achieve goal G is not saying that G is the efficient cause of B. On the contrary, B brings G into effect, if it is successful. If B is not successful, G never comes into being; if backwards causation was what was in question we would have here an effect without its cause.

At the other extreme, defenders of teleology have sometimes claimed that all causation is teleological. Causal laws must be stated in terms of the tendency of causal agents to produce certain effects unless interfered with. But are not laws stated in terms of tendencies teleological laws, since tendencies are defined in terms of their upshot? But an act may be defined by its result, and a tendency specified as a tendency to perform such an act, without the 'end' in the sense of final state being an 'end' in the sense of goal. A tendency is only teleological if it is a tendency to do something for the benefit of the agent, or of something bearing a special relation to the agent.

Any teleological explanation must involve an activity which can be done well or badly, or an entity for which there can be good or bad. The paradigm of such entities is the living organism: that has needs, can flourish, can sicken, decay, and die. There can be good or bad for things other than whole living organisms: things can be good or bad for the parts, artefacts, environments of living beings. But there are many items—numbers, classes, rocks, dust, mud, elementary particles and the like—for which there is no such thing as good and bad.

Once we have spelt out what is involved in the teleological phenomena which provide the basis for the argument from design, it is clear that the locus of that argument is the same as the locus of the problem of evil. It is the same kind of entity, the same realm of being, and the same features of that realm, that provide a home for the premises of both arguments. In order to specify what were the kinds of being to which teleological explanations were appropriate I had to bring in not only the notions of goodness, life, and flourishing, but also the notions of badness, decay, and death. Whatever can have things good for it can also have things bad for it.

This, as Farrer pointed out, is the first step towards the resolution of the problem of evil. The possibility of goodness brings with it the possibility of badness: if we can describe what is good for X we are *eo ipso* describing that whose lack is bad for X; if in saying that a particular X is a good X we are saying more than simply that it is an X, then there must be the possibility of describing an X which is not a good X. Whoever, therefore, makes a world in which there are things good for things, is making a world in which there is the logical possibility of things bad for things.

The problem of evil, of course, is a problem only for those who accept that there is a good creator, or at least a good ruler, of the Universe. If the world we see takes its origin and course from iron necessity, or blind chance, or some combination of the two, then evil may be regrettable but it is scarcely problematic: what reason have we to expect the world to be anything other than a vale of tears? Not even everyone who accepts the existence of an all-powerful creator need find the existence of evil logically disturbing. The first mover unmoved, the first cause of all, the *ens realissimum*, is not obviously, without considerable further argument, a source of goodness. It is the argument from design which leads to the conclusion that there is an extramundane origin precisely for purpose and the pursuit of good.

Consideration of the Argument from Design, therefore, is related to the resolution of the problem of evil in two different ways. If one rejects the argument, then one is, so far as natural theology is concerned, freed from the logical constraints of the problem of evil. If one accepts the argument, then one accepts along with it at least a partial recipe for the problem's solution: for the author of goodness to which the argument leads is by logical necessity the author of the possibility of evil.

This goes part of the way to the problem's solution: but must we not go much further? Must the logical possibility be actualized in the real world? Could not omnipotence make a world in which the possibility remained no more than a possibility?

The world we live in seems to have two features—emphasized by Farrer—which go beyond the necessity imposed by the nature of good and evil. In the first place, it is a world in which form survives by the precarious management of chaos: in which, for instance, my intellectual and animal life organizes the chemical and physical material in which it is embodied. Secondly, it is a world in which the organisms of various forms compete with each other for the matter to be organized: in which predators live off their prey and there is competition not only between but within species for the benefits offered by the environment.

A material world of precarious competition is the only world of which we have experience, and our imaginations are too feeble for us to be sure whether other forms of world are genuinely conceivable. The most sustained effort to imagine beings whose forms were not enthroned on chaos was the angelology of the medieval scholastics. It is difficult to be confident whether the immaterial spirits of scholastic tradition are genuinely conceivable or not; it is even more difficult to have much hope that we shall do better than the scholastics in this area in drawing limits to conceivability which are firm enough to rest an argument upon. We may, I think, accept with Farrer that a world containing any beings whom we could conceive of as having good would also have the actuality, and not just the possibility, of evil, because of the interlocking of one creature's good with another's evil. This is something which we must accept: as he puts it 'As we love our own distinct being, so must we endure the conditions of its possibility'.

Suppose we accept, then, that any world containing good must contain the possibility of evil, and accept that any world of the kind that has the likes of us in it must have the actuality of evil. Can one not still maintain that only a brute or blackguard would create the world we actually have? To consider this we have to take a further step in the consideration both of the problem of evil and of the argument from design.

Why is the presence of good and evil in the world supposed to call for an extra-terrestrial source? We don't, after all, think that the presence of hot and cold in the world means that there has to be an extra-terrestrial fount of heat: what is special about good and evil?

The argument from design turns on the fact that much of the good which is present in the world is present in the form of purpose. (I put on one side the question whether there could be a world in which there was good, but only accidental good; whether or not such a world is possible, ours is not such a world.) There are things which exist to serve purposes (e.g. organs with their distinct functions) and there are things which have purposes (e.g. animals with their characteristic activities).

I must avert a misunderstanding here: having a purpose does not involve, necessarily, knowledge or intention of that purpose. Not all purposes of entities are conscious goals or projects of that entity. The activity of the spider has as its purpose the construction of the web, as the activity of the dog has as its purpose the retrieval of a bone; but the dog is conscious of the purpose, as the spider is not. Not all purposeful actions are intentional actions, and not all entities with purposes are entities that have been designed by those whose needs they serve. Whether or not my liver was designed by God, it was not designed by me.

'Purpose', then, does not mean the same as 'design'. The argument from design aims to show that all purpose originates from design—but it does not assume this as if it was a tautology. Design is purpose which derives from a conception of the good which fulfils the purpose. If the conclusion of the argument from design is correct, then all purpose is of this kind. But that is not something to be assumed at the outset.

Nowadays, however, both proponents and critics of the Argument from Design accept the premise that naked purpose is inconceivable. That is to say, if we have an explanation in terms of purpose, it cannot be a fundamental, rock-bottom explanation. The explanation must be reducible to an explanation in terms of design, that is to say intelligent purpose; or to explanation of a mechanistic kind, in terms of necessity, chance, or both. Theists opt for the first kind of explanation; many evolutionary biologists for the second.

There are five levels at which prima facie there is purpose operative in the universe: first, the operation of mature living organisms; second, the operation of organs within those organisms; third, the morphogenesis of the individual from the embryonic state; fourth, the emergence of new species; fifth, the origin of speciation and of life itself. At each of these levels purpose may seem to call for a designer; at each level, one who wishes to resist this conclusion must reduce the teleological elements to mechanistic ones, claiming to show how the evolution of life is either an inevitable process, explained by the natural properties of non-living matter, or the result of the operation of necessitating forces upon the outcome of chance occurrences.

I shall not consider in detail the plausibility or otherwise of mechanistic reduction of teleology at each of these five levels. I shall assume, for the sake of argument, that at one or other point the reduction breaks down, so that the argument from design succeeds. I ask, what consequences follow for the problem of evil and the responsibility of the cosmic designer?

The answer seems to differ according to the method by which the design operates or, if you like, the point or points at which the purposiveness is introduced from outside into the cosmic story. The first case, and the easiest one to judge, is the one in which the designer achieves his purpose, and the purposes of his creatures, by the operation of necessitating laws. In such a world, it seems, God would be not only the author of evil, but the author of sin. As I put it in my book *The God of the Philosophers*:

> If an agent freely and knowingly sets in motion a deterministic process with a certain upshot, it seems that he must be responsible for that upshot. Calvin argued rightly that the truth of determinism would not make everything that happens in the world happen by God's intention:

only some of the events of history would be chosen by God as ends or means, others could be merely consequences of his choices. But that would not suffice to acquit God of responsibility for sin. For moral agents are responsible not only for their intentional actions, but also for the consequences of their actions: for states of affairs which they bring about voluntarily but not intentionally. An indeterminist can make a distinction between those states of affairs which God causes, and those which he merely permits: but in a deterministic created universe, the distinction between causing and permitting would have no application to God. (p. 86)

This consideration is unlikely greatly to trouble a proponent of the argument from design, since our universe does not appear to be one in which determinism reigns, but rather one in which, while there are effects which are determined by causes, there are also events which are determined only by coming to pass. Indeed the mechanistic opponents of the argument from design themselves commonly seek to reduce purpose not to determinist necessity, but to the operation of necessity upon chance events.

We must look more closely at what we mean by chance, considered as an explanatory factor. One kind of chance is the chance which is the unsought outcome of the operation of one or more causes (where more than one cause is in play, this kind of chance is coincidence). The other kind of chance is the tendency to produce its proper effect *n* times out of *m*. The two kinds may be linked together in a particular case: a throw of a double six when dicing is an instance of both kinds of chance. Chance in the second sense is a genuine—if indeterministic—principle of explanation.

Freedom is not the same thing as chance. An action is free if it is the exercise of a voluntary power; and a voluntary power differs from a natural power in being a two-way power. The notion of chance applies to voluntary powers no less than to natural power. Just as one kind of chance consists in the coincidental exercise of the natural powers of unrelated agents, so another kind of chance consists in the coincidental operation of non-conspiring voluntary causes.

It is often claimed that allowing the reality of freedom and chance in our world is the key to resolving the problem of evil. This, I believe, is not so.

First, if compatibilism is true, as I have argued on several occasions elsewhere, then the acknowledgement of freedom does not even rule out the possibility of the deterministic universe in which God would undoubtedly be the author of sin.

Secondly, the kinds of chance we have recognized are compatible both with design and with the responsibility of the designer. A designer may put together two non-conspiring causes in such a way that the outcome is one not sought (pursued, tended towards) by either cause; he may include among the causes indeterministic ones (as a computer programmer may include a randomizing element in his program). In neither case would he avoid responsibility for what happens, despite the attempt to show the contrary by Descartes in his celebrated parable of the King who both forbids duelling but brings two inveterate duellers together in a quarrel. (AT iv, 35)

What of undesigned chance: will that absolve the maker of the world for responsibility for the evils it contains? Evils which are the consequences of undesigned chance would be neither means nor ends of the Great Designer. They would be risks which he takes—knowingly, in general, of the nature of the risk; but without knowledge, in particular, of the evils which will in fact eventuate. A designer who takes risks of this kind would be less, I have argued elsewhere, than the God of traditional Western theism, because he would not have full knowledge of the future. But our present question is: would he avoid responsibility for the evils of the world?

The natural response is to say that it all depends whether the game is worth the candle: whether the goods to be achieved are worth the risk of the evil. If this is so, then only a global view of the totality of good and evil to be found in the achieved universe would enable one to cast the accounts. And this means that no impugner of divine goodness could hope to make his prosecution succeed: for the evidence which could alone secure a conviction is available only to the accused, and not even to him in advance of the end of the cosmic drama.

Note that a theist could adopt this response to the Problem of Evil without taking the view that no moral judgement is possible about God. For if he is, as he is likely to be, an absolutist in morals, he will agree that there are certain things God could not do and remain good: such as telling lies, or punishing the innocent everlastingly. He will not need to adopt the consequentialist view that moral judgement on an action—whether a human action or a divine action—must wait on a full conspectus of the consequences. It is only in the case where evil is risked—not when it is knowingly permitted or wilfully brought about—that the felicific calculus is allowed to have moral weight.

But this kind of reflection brings about the unreality of the exercise we have been engaged in. It must be doubtful whether cosmic judgements of the form 'the world is on balance a good/bad place' have any clear sense; anyone who believes they have, must believe that the sense of a judgement is totally divorced from the possibility of the judger's putting himself into a position to have adequate grounds for the judgement. It is hard enough to attach sense to much more modest generalizations such as 'the human race is on the whole a good/bad thing' or 'people in the twentieth century are happier/unhappier than people were in the twelfth century'.

If it is difficult to attach clear sense to the evidence to be brought against the designer of the world, it is even more difficult to take seriously the idea of calling him before the bar of human morality. Morality presupposes a moral community: and a moral community must be of beings with a common language, roughly equal powers, and roughly similar needs, desires and interests. God can no more be part of a moral community with human beings than he can be part of a political community with them. As Aristotle said, we cannot attribute moral virtues to divinity: the praise would be vulgar. Equally, moral blame would be laughable.

Remember that we have been speaking throughout within the bounds of natural theology. If an alleged revelation claims that God has entered into moral relationships with human beings, then we enter into a different realm of discourse: but if that discourse can be made intelligible, the present difficulty is one that will have to be surmounted along with others. Within the realm of a purely natural theology, there is no problem of evil; but equally we must retract the claim that the argument from design showed God to be good.

Farrer was right to say that 'I believe in God because the world is so bad' is as sound an argument as 'I believe in God because the world is so good'. But he did not follow sufficiently rigorously his own insight that the arguments for the existence of God start from a division within the finite and show that that insight is transcended in the infinite. Farrer was right to show that the argument from design and the problem of evil are two formulations of a single progress from the finite dichotomy of good and evil to an infinite in which that dichotomy is transcended. But that progress leads to a God which is no more the source of good than the source of evil. The God to which this argument of rational theology leads is not supreme goodness: it is a being which is beyond good and evil.

# 7 John Henry Newman on the Justification of Faith

John Henry Newman's major contribution to philosophy was his *Essay in Aid of a Grammar of Assent*, published in 1870. This book centres upon a question of primary importance in the philosophy of religion: how can religious belief be justified, given that the evidence for its conclusions seems so inadequate to the degree of its commitment? The book contains much original material of interest on many philosophical topics. But on the precise question of the nature and justification of faith some of Newman's very best work occurs not here but in his earlier University Sermons, whose full title is *Sermons chiefly on the theory of religious belief, preached before the University of Oxford.* These sermons were preached between 1826 and 1843, between Newman's appointment as a college tutor in Oxford and his resignation of the living of the University Chuch of St Mary's, all of them while he was a Fellow of Oriel. There is no great difference in actual doctrine between Newman's Anglican and Catholic writings on this topic, and where there are differences they seem not to depend on religious or doctrinal grounds. There are at least as great differences between his earlier and later Oriel sermons as between the late Oriel sermons and the *Grammar*.

In the theological tradition in which Newman wrote faith was contrasted on the one hand with reason and knowledge and on the other with hope and charity. 'Faith' was used in a narrower sense than 'belief'. Aristotle believed that there was a divine prime mover unmoved; but his belief was not faith in God. On the other hand, Marlowe's Faustus, on the verge of damnation, speaks of Christ's blood streaming in the firmament; he has lost hope and charity yet retains faith. So faith contrasts both with reason and with love. The special nature of the belief that is faith is that it is a belief in something as revealed by God; belief in a proposition on the word of God.

This is a Catholic, not a Protestant view of the naure of faith. Newman held it already in his University Sermons.

> The Word of Life is offered to a man; and on its being offered, he has Faith in it. Why? On these two grounds—the word of its human messenger, and the likelihood of the message. And why does he feel the message to be probable? Because he has a love for it, his love being strong, though the testimony is weak. He has a keen sense of the intrinsic excellence of the message, of its desirableness, of its likeness to what it seems to him Divine goodness would vouchsafe did He vouchsafe any. (U, 195)

Newman attacks the idea that reason judges both the evidence for and the content of revelation, and opposes the view that faith is just a state of heart, a moral quality, of adoration and obedience. Faith is itself an intellectural quality, even though reason is not an indispensable preliminary to faith. (U, 173)

What is the role of reason? We have direct knowledge of material things through the senses: we are sensible of the existence of persons and things, we are directly cognizant of them through the senses. (To think that we have faculties for direct knowledge of immaterial things is a form of enthusiasm; certainly we are not conscious of any such faculties.) The senses are the only instruments which we know to be granted to us for direct and immediate acquaintance with things external to us. Even our senses convey us but a little way out of ourselves: we have to be near things to touch them: we can neither see hear nor touch things past or future. (U, 197–8)

> Now reason is that faculty of the mind by which this deficiency is supplied; by which knowledge of things external to us, of beings, facts, and events, is attained beyond the range of sense. It ascertains for us not natural things only, or immaterial only, or present only, or past or future; but, even if limited in its power, it is unlimited in its range . . . It reaches to the ends of the universe, and to the throne of God beyond them; it

> brings us knowledge, whether clear or uncertain, still knowledge, in whatever degree of perfection, from every side; but, at the same time, with this characteristic that it obtains it indirectly, not directly. (U, 199)

Reason does not really perceive any thing; but is a faculty of proceeding from things that are perceived to things which are not. It is the faculty of gaining knowledge upon grounds given; and its exercise lies in asserting one thing, because of some other thing. When its exercise is conducted rightly, it leads to knowledge; when wrongly, to apparent knowledge, to opinion, and error. (U, 199)

If this be reason, then faith, simply considered, is itself an exercise of reason, whether right or wrong. For example, 'I assent to this doctrine as true, because I have been taught it', or 'because persons whom I trust say it was once guaranteed by miracles'. It 'must be allowed on all hands' says Newman 'either that [faith] is illogical, or that the mind has some grounds which are not fully brought out when the process is thus exhibited.' The world says faith is weak; Scripture says it is unearthly. (U, 200–1) Faith is an act of Reason, but of what the world would call weak, bad, or insufficient reason, and that because it rests on presumption more and on evidence less.

Newman says it is true that nothing is true or right but what may be justified and in a certain sense proved by reason. But that does not mean that faith is grounded on reason; unless a judge can be called the origin as well as the justifier of the innocence of those who are brought before him. (U, 174) On a popular view, reason requires strong evidence before assent, faith is content with weaker evidence. So Hume, Bentham, and all those who like them think that faith is credulity. But in fact credulity is the counterfeit of faith, as scepticism is of reason. (U, 177)

> Faith . . . does not demand evidence so strong as is necessary for . . . belief on the ground of Reason; and why? For this reason, because it is mainly swayed by antecedent considerations . . . previous notices, prepossessions, and (in a good sense of the word) prejudices. The mind that believes is acted upon by its own hopes, fears, and existing opinions . . . previously entertained principles, views, and wishes. (U, 179–80)

Unbelievers say that a man is as little responsible for his faith as for his bodily functions; both are from nature, and the will cannot make a weak proof a strong one.

But love of the great Object of Faith, watchful attention to Him, readiness to believe Him near, easiness to believe Him interposing in human affairs, fear of the risk of slighting or missing what may really have come from Him; these are feelings not natural to fallen man, and they come only of supernatural grace; and these are the feelings which make us think evidence sufficient, which falls short of a proof in itself. (U, 185)

Thus we can see how Faith is and is not according to Reason: taken together with the antecedent probability that Providence will reveal himself, otherwise deficient evidence may be enough for conviction, even in the judgement of Reason.

That is, Reason, weighing evidence only, or arguing from external experience, is counter to Faith; but, admitting the full influence of the moral feelings, it concurs with it. (U, 187)

*De facto* this was how it all happened in the preaching of Christ and the apostles. It is wrong to think oneself a judge of religious truth without preparation of heart.

Gross eyes see not; heavy ears hear not. But in the schools of the world the ways towards Truth are considered high roads open to all men, however disposed, at all times. Truth is to be approached without homage. Every one is considered on a level with his neighbour; or rather, the powers of the intellect, acuteness, sagacity, subtlety and depth, are thought the guides into Truth. Men consider that they have as full a right to discuss religious subjects, as if they were themselves religious. They will enter upon the most sacred points of Faith at the moment, at their pleasure—if it so happen, in a careless frame of mind, in their hours of recreation, over the wine cup. Is it wonderful that they so frequently end in becoming indifferentists? (U, 190–1)

The mismatch between evidence and commitment, and the importance of previous attitudes, is to be observed not only in religious faith, but in other cases of belief.

We hear a report in the streets, or read it in the public journals. We know nothing of the evidence; we do not know the witnesses, or anything about them: yet sometimes we believe implicitly, sometimes not; sometimes we believe without asking for evidence, sometimes we disbelieve till we receive it. Did a rumour circulate of a destructive earthquake in Syria or the South of Europe, we should readily credit it; both because it might easily be true, and because it was nothing to us though it were. Did the report relate to countries nearer home, we should try to trace and authenticate it. We do not call for evidence till antecedent probabilities fail. (U, 180)

Newman goes on to develop the theme that Faith is not the only exercise of reason which, when critically examined, would be called unreasonable and yet is not so. Choice of sides in political questions, decisions for or against economic policies, tastes in literature: in all such cases if we measure people's grounds merely by the reasons they produce we have no difficulty in holding them up to ridicule, or even censure. So too with prophecies of weather, judgements of character, and even theories of the physical world. (U, 202)

However systematically we argue on any topic, there must ever be something assumed ultimately which is incapable of proof, and without which our conclusion will be as illogical as faith is apt to seem to men of the world. We trust our senses without proof; we rely implicitly on our memory, and that too in spite of its being obviously unstable and treacherous. We trust to memory for the truth of most of our opinions; the grounds on which we hold them not being at a given moment all present to our minds

It may be said that without such assumption the world could not go on: true, and in the same way the Church could not go on without Faith. Acquiescence in testimony, or in evidence not stronger than testimony, is the only method, so far as we see, by which the next world can be revealed to us. (U, 206–7)

Moreover, the more precious a piece of knowledge is, the more subtle the evidence on which it is received.

We are so constituted that if we insist upon being as sure as is conceivable, in every step of our course, we must be content to creep along the ground, and can never soar. If we are intended for great ends, we are called to great hazards; and whereas we are given absolute certainty in nothing, we must in all things choose between doubt and inactivity. (U, 208)

In the pursuit of power, distinction in experimental science, or character for greatness, we cannot avoid risk. Great objects exact a venture and sacrifice is the condition of honour; so

even though the feelings which prompt us to see God in all things, and to recognize supernatural works in matters of the world, mislead us at times, though they make us trust in evidence which we ought not to admit, and partially incur with justice the imputation of credulity, yet a Faith which generously apprehends Eternal truth, though at times it degenerates into superstition, is far better than that cold, sceptical, critical tone of mind, which has no inward sense of an overrulling, ever-present Providence, no desire to approach its God, but sits at home waiting for the fearful clearness of his visible coming, whom it might seek and find in due measure amid the twilight of the present world. (U, 213)

The mind ranges to and fro, and spreads out, and advances forward with a quickness which has become a proverb and a subtlety and versatility which baffle investigation. It passes on from point to point, gaining one by some indication, another on a probability; then availing itself of an association; then falling back on some received law; next seizing on testimony; then committing itself to some popular impression, or some inward instinct, or some obscure memory; and thus it makes progress not unlike a clamberer on a steep cliff, who, by quick eye, prompt hand, and firm foot, ascends how he knows not himself, by personal endowments and by practice, rather than by rule, leaving no track behind him, and unable to teach another. It is not too much to say that the stepping by which great geniuses scale the mountains for truth is as unsafe and precarious to men in general as the ascent of a skilful mountaineer up a literal crag. It is a way which they alone can take; and its justification lies in its success (U, 252–3)

But how can one tell what is success in religious matters? On Newman's own account, there is a close similarity between faith and bigotry. In each case the grounds are conjectural, the issue is absolute acceptance of a certain message or doctrine as divine. Faith 'starts from probability, yet it ends in peremptory statements, if so be, mysterious, or at least beyond experience. It believes an informant amid doubt, yet accepts his information without doubt.'

The University Sermons do not really succeed in solving the problem, to which Newman returned in the *Grammar of Assent*. How is it that a proposition which is not, and cannot be, demonstrated, which at the highest can only be proved to be truth-like, not true, nevertheless claims and receives our unqualified adhesion?

Some philosophers, for example Locke, say that there can be no demonstrable truth in concrete matter, and therefore assent to a concrete proposition must be conditional. Probable reasoning can never lead to certitude. According to Locke, there are degrees of assent, and absolute assent has no legitimate exercise except as ratifying acts of intuition or demonstration.

Locke gives, as the unerring mark of the love of truth, the not entertaining any proposition with greater assurance than the proofs it is built on will warrant. 'Whoever goes beyond this measure of assent, it is plain, receives not truth in the love of it, loves not truth for truth-sake, but for some other by-end' (*Essay on Human Understanding*, IV, xvi, 6).

This doctrine of Locke's is one of Newman's main targets of attack. In *The Development of Doctrine* he says that the by-end may be the love of God. (ch. vii. 2) In the *Grammar of Assent* he claims that Locke's thesis is insufficiently empirical, too idealistic. Locke calls men 'irrational and indefensible if (so to speak) they take to the water, instead of remaining under the narrow wings of his own arbitrary theory'.

On Locke's view, says Newman, assent would simply be a mere reduplication or echo of inference, 'assent' just another name for inference. But in fact the two do not always go together; one may be strong and the other weak. We often assent, when we have forgotten the reasons for our assent. Reasons may still seem strong, and yet we do not any longer assent. Sometimes assent is never given in spite of strong and convincing arguments, perhaps through prejudice, perhaps through tardiness. Arguments may be better or worse, but assent either exists or not. (G, 110–12)

Even in mathematics there is a difference between inference and assent. A mathematician would not assent to his own conclusions, on new and difficult ground, and in the case of abstruse calculations, however often he went over his work, until he had the corroboration of other judgements besides his own. (G, 113)

In demonstrative matters assent excludes doubt. In concrete cases, we do not give doubtful assent, for are there instances where we assent a little and not much.

Usually we do not assent at all. Every day, as it comes, brings with it opportunities for us to enlarge our circle of assents. We read the newspapers; we look through debates in Parliament, pleadings in the law courts, leading articles, letters of correspondents, reviews of books, criticism in the fine arts, and we either form no opinion at all upon the subjects discussed, as lying out of our line, or at most we have only an opinion about them . . . we never say that we give [a proposition] a degree of assent. We might as well talk of degrees of truth as degrees of assent. (G, 115)

But there are unconditional assents on evidence short of intuition and demonstration. We all believe without any doubt that we exist; that we have an individuality and identity all our own; that we think, feel, and act, in the home of our own minds.

Nor is the assent which we give to facts limited to the range of self-consciousness. We are sure beyond all hazard of a mistake, that our own self is not the only being existing; that there is an external world; that it is a system with parts and a whole, a universe carried on by laws; and that the future is affected by the past. We accept and hold with an unqualified assent, that the earth, considered as a phenonemon, is a globe; that all its regions see the sun by turns; that there are vast tracts on it of land and water; that there are really existing cities on definite sites, which go by the names of London, Paris, Florence and Madrid. We are sure that Paris or London, unless suddenly swallowed up by an earthquake or burned to the ground, is today just what it was yesterday, when we left it. (G, 117)

Newman's favourite example of a firm belief on flimsy evidence is our conviction that Great Britain is an island.

Our reasons for believing that we are circumnavigable are such as these: first, we have been so taught in our childhood, and it is so in all the maps; next, we never heard it contradicted or questioned; on the contrary, every one whom we have heard speak on the subject of Great Britain, every book we have read, invariably took it for granted; our whole national history, the routine transactions and current events of the country, our social and commercial system, our political relations with foreigners, imply it in one way or another. Numberless facts, or what we consider facts, rest on the truth of it; no received fact rests on its being otherwise . . .

However, negative arguments and circumstantial evidence are not all, in such a matter, which we have a right to require. They are not the highest kind of proof possible. Those who have circumnavigated the island have a right to be certain: have we ever ourselves fallen in with anyone who has? . . . Have we personally more than an impression, if we view the matter argumentatively, a lifelong impression about Great Britain, like the belief, so long and so widely entertained, that the earth was immovable, and the sun careered round it? I am not at all insinuating that we are not rational in our certitude; I only mean that we cannot analyse a proof satisfactorily, the result of which good sense actually guarantees to us. (G, 191–2)

Take another example.

What are my grounds for thinking that I, in my own particular case shall die? I am as certain of it in my own innermost mind, as I am that I now live; but what is the distinct evidence on which I allow myself to be certain? How would it tell in a court of justice?. . . Men tell me that there is a law of death, meaning by a law a necessity; and I answer that they are throwing dust into my eyes, giving me words instead of things. What is a law but a generalized fact? And what power has the past over the future? and what power has the case of others over my own case? and how many deaths have I seen? How many ocular witnesses have imparted to me their experience of deaths, sufficient to establish what is called a law?. . .

The strongest proof I have for my inevitable mortality is the reductio ad absurdum. Can I point to the man, in historic times, who has lived his two hundred years? What has become of past generations of men, unless it is true that they suffered dissolution? But this is a circuitous argument to warrant a conclusion to which in matter of fact I adhere so relentlessly. (G, 195)

We laugh to scorn the idea that we had no parents though we have no memory of our birth; that we shall never depart this life, though we can have no experience of the future; that we are able to live without food, though we have never tried; that a world of men did not live before our time, or that that world has no history; that there has been no rise and fall of states, no great men, no wars, no revolutions, no art, no science, no literature, no religion. (G, 117)

On all these truths, Newman sums up, we have an immediate and unhesitating hold,

> nor do we think ourselves guilty of not loving truth for truth's sake, because we cannot reach them through a series of intuitive propositions. . . None of us can think or act without the acceptance of truths, not intuitive, not demonstrated, yet sovereign. (G, 118)

Philosophers like Locke do not really have misgivings about the truths they call in question;

> they think it a duty to remind us that since the full etiquette of logical requirements has not been satisfied, we must believe those truths at our peril. . . . They do not, for instance, intend for a moment to imply that there is even the shadow of a doubt that Great Britain is an island, but they think we ought to know, if we do not know, that there is no proof of the fact, in mode and figure, equal to the proof of a proposition of Euclid; and that in consequence they and we are all bound to suspend our judgement about such a fact, though it be in an infinitesimal degree, lest we should seem not to love truth for truth's sake. Having made their protest, they subside without scruple into that same absolute assurance of only partially proved truths, which is natural to the illogical imagination of the multitude. (G, 119)

Newman makes a distinction between simple assent and complex assent. Simple assent is often unconscious. There are innumerable acts of assent which we make without reflecion. But complex, or reflex, assent is what is meant by certitude: and it is certitude that is the characteristic manifestation of religious faith. Newman describes certitude in the following way:

> It seems then that on the whole there are three conditions of certitude: that it follows on investigation and proof, that it is accompanied by a specific sense of intellectual satisfaction and repose, and that it is irreversible. If the assent is made without rational grounds, it is a rash judgement, a fancy, or a prejudice; if without the sense of finality, it is scarcely more than an inference; if without permanence, it is a mere conviction. (G, 168)

But how can faith be certitude, if certitude follows on investigation? Does not investigation imply doubt, which conflicts with faith? To set about concluding a proposition is not *ipso facto* to doubt its truth: we may aim at inferring a proposition, while all the time we assent to it; we do not deny our faith because we become controversialists. Investigation is not inquiry; inquiry is indeed inconsistent with assent. It is sometimes complained of that a Catholic cannot inquire into the truth of his creed: of course he cannot if he would retain the name of believer. (G, 125)

But may not investigation lead to giving up assent? Yes, it may; but

> my vague consciousness of the possibility of a reversal of my belief in the course of my researches, as little interferes with the honesty and firmness of that belief while those researches proceed, as the recognition of the possibility of my train's oversetting is an evidence of an intention on my part of undergoing so great a calamity (G, 127)

Newman describes the specific feeling of certainty:

> a feeling of satisfaction and self-gratulation. The repose in self and in its object, as connected with self, which is characteristic of Certitude, does not attach to mere knowing, that is, to the perception of things, but to the consciousness of having that knowledge. (G, 134)

Assents may and do change; certitudes endure. This is why religion demands more than an assent to its truth; it requires a certitude, or at least an assent which is convertible into certitude on demand. Belief does not necessarily imply a positive resolution in the party believing never to abandon the belief. It implies not an intention never to change, but the utter absence of all thought, or expectation or fear of change.

Newman from time to time talks as if there is such a thing as false certitude, a state which differs from knowledge only in its truth value. But, he says, not altogether consistently, if the proposition is objectively true, 'then the assent may be called a perception, the conviction a certitude, the proposition or truth a certainty, or thing known, or a matter of knowledge, and to assent to it is to know'. (G, 128)

Whether or not certitude entails truth, it is undeniable that to be certain of something involves believing in its truth. It follows that if I am certain of a thing, I believe it will remain what I now hold it to be, even though my mind should have the bad fortune to let it drop. If we are certain, we spontaneously reject objections to our belief as idle; though the contradictory of a truth be brought back to mind by the pertinacity of an opponent, or a voluntary or involuntary act of imagination, still that contradictory proposition and its arguments are mere phantoms and dreams. This is like the way the mind revolts from the supposition that a straight line is the longest distance between two points, or that Great Britain is in shape an exact square, or that I shall escape dying. (G. 130)

Some may say, we should never have this contempt-bringing conviction of anything; but if in fact 'a man has such a conviction, if he is sure that Ireland is to the West of England, or that the Pope is the Vicar of Christ, nothing is left to him, if he would be consistent, but to carry his conviction out into this magisterial intolerance of any contrary assertion'. Newman goes on to say: Whoever loses his conviction on a given point is thereby proved not to have been certain of it. (G, 130 f.)

But is there any specific state or habit of thought, of which the distinguishing mark is unchangeableness? On the contrary, any conviction, false as well as true, may last; and any conviction, true as well as false, may be lost. No line can be drawn between such real certitudes as have truth for their object, and apparent certitudes. There is no test of genuine certitude of truth. What looks like certitude always is exposed to the chance of turning out to be a mistake. Certitude does not admit of an interior, immediate test, sufficient to discriminate it from false certitude. (G, 145)

Newman correctly distinguishes certainty from infallibility. My memory is not infallible; I remember for certain what I did yesterday, but that does not mean that my memory is infallible. I am quite clear that two and two make four, but I often make mistakes in long addition sums. Certitude concerns a particular proposition, infallibility is a faculty or gift. It is possible to be certain that Victoria is Queen, without claiming infallibility, as it is possible to do a virtuous action without being impeccable. (G, 147)

But how can the secure repose of certitude be mine if I know, as I know too well, that before now I have thought myself certain when I was certain after all of an untruth? What happened once may happen again. Newman's answer is this: mistakes should make us more cautious, but even so grounds for caution may be overcome.

Suppose I am walking out in the moonlight, and see dimly the outlines of some figure among the trees;—it is a man. I draw nearer, it is still a man; nearer still, and all hesitation is at an end,—I am certain it is a man. But he neither moves nor speaks when I address him; and then I ask myself what can be his purpose in hiding among the trees at such an hour. I come quite close to him and put out my arm. Then I find for certain that what I took for a man is but a singular shadow, formed by the falling of the moonlight on the interstices of some branches or their foliage. Am I not to indulge my second certitude, because I was wrong in my first? Does not any objection, which lies against my second from the failure of my first, fade away before the evidence on which my second is founded? (G, 151)

We do not dispense with clocks, because from time to time they go wrong and tell untruly.

The sense of certitude may be called the bell of the intellect; and that it strikes when it should not is a proof that the clock is out of order, no proof that the bell will be untrustworthy and useless when it comes to us adjusted and regulated from the hands of the clockmaker (G, 152)

Certitude is a mental state; certainty is a quality of propositions. Those propositions I call certain, which are such that I am certain of them. Certitude is . . . an active recognition of propositions as true, such as it is the duty of each individual himself to exercise at the bidding of reason, and, when reason forbids, to withold. . . . And reason never bids us be certain except on an absolute proof; and such a proof can never be furnished to us by the logic of words, for as certitude is of the mind, so is the act of inference which leads to it.

Is there any criterion of the accuracy of an inference?

the sole and final judgement on the validity of an inference in concrete matter is committed to the personal action of the ratiocinative faculty, the perfection or virtue of which I have called the illative sense, a use of the word sense parallel to our use of it in 'good sense' 'commmon sense' . . . (G, 223)

We have to aceept being the kind of things we are: beings which have to progress by inference and assent. The course of inference is ever more or less obscure, while assent is ever distinct and definite, yet one follows on the other; we have to accept this. But the illative sense is in theoretical reasoning what Aristotle's *phronesis* is in practical reasoning. Aristotle says that no code of laws, or moral treatise, maps out the path of individual virtue. So too with the controlling principle in inferences. There are as many forms of *phronesis* as there are virtues. There is no one formula which is a working rule for poetry, medicine, politics; so too with ratiocination. In reasoning on any subject whatever, which is concrete, we proceed, as far indeed as we can, by the logic of language; but we are obliged to supplement it by the more subtle and elastic logic of thought.

How does Newman apply this to the evidences for religion? Christianity is a revelation, a

> definite message from God to man distinctly conveyed by his chosen instruments, and to be received as such a message; and therefore to be positively acknowledged, embraced, and maintained as true, on the grounds of its being divine, not as true on intrinsic grounds, not as probably true, or partially true, but as absolutely certain knowledge, certain in a sense in which nothing else can be certain, because it comes from Him who neither can deceive nor be deceived. (G, 250)

With regard to the justification of religious belief, Newman gives up the intention of demonstrating either natural religion or Christianity.

> Not that I deny that demonstration is possible. Truth, certainly, as such, rests upon grounds intrinsically and objectively and abstractedly demonstrative, but it does not follow from this that the arguments producible in its favour are unanswerable and irresistible. . . . The fact of revelation is in itself demonstrably true, but it is not therefore true irresistibly; else how comes it to be resisted? (G, 264)

'For me' says Newman 'it is more congenial to my own judgement to attempt to prove Christianity in the same informal way in which I can prove for certain that I have been born into this world, and that I shall die out of it.' (G, 264)

Newman's proof will work only for those who are prepared for it, imbued with religious opinions and sentiments indentified with natural religion. He assumes the falsehood of the opinions which 'characterize a civilized age'. The evidences 'presuppose a belief and perception of the Divine Presence'. Newman does not stress miracles, but rather 'those coincidences and their cumulations which, though not in themselves miraculous, do irresistibly force upon us, almost by the law of our nature, the presence of the extraordinary agency of Him whose being we already acknowledge.'

As example Newman quotes the sudden death of a market woman following the utterance of a curse, and the fact of Napoleon's being defeated in Russia within two years of his being excommunicated by the Pope. These coincidences are indications, to the illative sense of those who believe in a Moral Governor, of his immediate presence. But the greatest of these impressive coincidences is the whole history of Judaism and Christianity.

If the history of Judaism is so wonderful as to suggest the presence of some special divine agency in its appointments and fortunes, still more wonderful and divine is the history of Christianity; and again it is more wonderful still, that two such wonderful creations should span almost the whole course of ages, during which nations and states have been in existence, and should constitute a professed system of continued intercourse between earth and heaven from first to last amid all the vicissitudes of human affairs. This phenomenon again carries on its face, to those who believe in a God, the probability that it has that divine origin which it professes to have (G, 283)

Newman concludes:

Christianity is addressed, both as regards its evidences and its contents, to minds which are in the normal condition of human nature, as believing in God and in a future judgement. Such minds it addresses both through the intellect and through the imagination; creating a certitude of its truth by arguments too various for direct enumeration, too personal and deep for words, too powerful and concurrent for refutation. Nor need reason come first and faith second (though this is the logical order) but one and the same teaching is in different aspects both object and proof, and elicits one complex act both of inference and assent. (G, 316)

Given Newman's own description of the scope of his argument, one may ask: Why should one believe in God and in a future judgement at all? In response to this question Newman makes his celebrated appeal to the testimony of conscience. He is not confident in the probative force of the traditional arguments to the existence of God from the nature of the physical world.

It is indeed a great question whether Atheism is not as philosophically consistent with the phenomena of the physical world, taken by themselves, as the doctrine of a creative and governing Power. But, however this be, the practical safeguard against Atheism in the case of scientific inquirers is the inward need and desire, the inward experience of that Power, existing in the mind before and independently of their examination of His material world (U, 186)

As from a multitude of instinctive perceptions, acting in particular instances, of something beyond the senses, we generalise the notion of an external world, and then picture that world in and according to those particular phenomena from which we started, so from the perceptive power which identifies the intimations of conscience with the reverberations or echoes (so to say) of an external admonition, we proceed on the notion of a Supreme Ruler and Judge. (G, 72)

Conscience is a mental phenomenon as much as memory, reason, or the sense of the beautiful. It is a moral sense and a sense of duty; a judgement of the reason and a magisterial dictate, it has both a critical and judicial office. Conscience, considered as a moral sense, is an intellectual sentiment, but it is always emotional; therefore it involves recognition of a living object. Inanimate things cannot stir our affections, these are correlative with persons.

If, on doing wrong, we feel the same tearful, broken hearted sorrow which overwhelms us on hurting a mother; if on doing right, we enjoy the same sunny serenity of mind, the same soothing, satisfactory delight which follows on our receiving praise from a father, we certainly have within us the image of some person, to whom our love and veneration look, in whose smile we find our happiness, for whom we yearn, towards whom we direct our pleadings, in whose anger we are troubled and waste away. These feelings in us are such as require for their exciting cause an intelligent being. . . . (G, 76)

So far I have expounded Newman, without criticizing him. I wish to end by stating briefly my own position on the issues on which he wrote so eloquently.

Newman begins his own criticism of Locke with the following words: 'I have so high a respect both for the character and the ability of Locke . . . that I feel no pleasure in considering him in the light of an opponent.' (G, 107) The Oxford philosopher H. H. Price, writing on the topic of belief, said

> Let us follow this excellent example; for no one, and certainly no Oxford man, should criticize Newman without praising him. . . . Newman is one of the masters of English prose. The power, and the charm, of his style are so compelling that the reader soon becomes their willing captive, and it seems ungrateful, almost ungracious, to question what has been so felicitously said. (*Belief*, 133)

One's reluctance to take a stand against Newman is increased by the fact that Newman puts the objections to his own views so marvellously well: indeed, he is often at this best when stating a position against which he intends to argue. Let us admire, for instance, the way in which he states the argument which is most likely to have occurred to those who have followed his defence of the justification of Christian belief.

> Antecedent probabilities may be equally available for what is true and what pretends to be true, for a revelation and its counterfeit, for Paganism, or Mahometanism, or Christianity. They seem to supply no intelligible rule what is to be believed and what not; or how a man is to pass from a false belief to a true. If a claim of miracles is to be acknowledged because it happens to be advanced, why not for the miracles of India as well as for those of Palestine? If the abstract probability of a Revelation be the measure of genuineness in a given case, why not in the case of Mahomet as well as of the Apostles? (U, 226)

The argument against Newman's position here could hardly be better put; and so it is in many other cases where Newman maintains implausible and contentious opinions. None the less, I cannot conclude without stating that Newman's account of the nature and justification of faith is wrong on a number of major points. I will list, without defending, five criticisms which can be made of his position.

1. First, despite what Newman says, assent does have degrees and this is true in religious matters as in others. This is something which Newman himself knows and admits when he is off his guard. There is a difference between an assent to a proposition without fear of its falsehood but with a readiness to examine contrary evidence and change one's mind, and an assent like Newman's certitude which contemns all objections which may be brought against it. Newman himself gives examples of adherence to propositions which does not fulfil the conditions of certitude. Some of these concern matters of religious belief.

> I may believe in the liquefaction of St Pantaleon's blood, and believe it to the best of my judgement to be a miracle, yet supposing a chemist offered to produce exactly the same phenomena under exactly similar circumstances by the materials put at his command by his science, so as to reduce what seemed beyond nature within natural laws, I should watch with some suspense of mind and misgiving the course of his experiment, as having no Divine Word to fall back upon as a ground of certainty that the liquefaction was miraculous. (G, 132)

This is a very important passage, which gives away Newman's official position. It shows that there is such a thing as belief, and indeed religious belief, which falls short of unconditional assent. The real question which Newman ought to be facing, is this: why is not this kind of certitude the appropriate kind in religious matters, given the nature of the evidence for there being a divine revlation of Christianity?

2. Newman is right to emphasize, and it is one of his major contributions to philosophy, that a belief such as the belief that Great Britain is an island is not a belief based on sufficient evidence. But the reason for this is that it is not based on evidence at all. For evidence has to be better known than that for which it is evidence; and none of the scraps of reasons I could produce for the proposition that Great Britain is an island are better known than the proposition itself.

But this means that there is not the parallel which Newman

drew between the belief that Great Britain is an island and the religious faith of a Christian believer. For faith to be faith and not mere belief it has to be belief on the word of God. If that is so, then the fact of revelation has to be better known than the content of revelation. But this Newman does not even attempt to prove.

3. Again, Newman is quite unconvincing in claiming that certitude is indefectible. It is true that knowledge is indefectible: If I claim to know that *p*, and then change my mind about *p*, I also withdraw the claim that I ever knew that *p*. But certainty is not like knowledge here: there is nothing odd in saying 'I was certain but I was wrong'. The difference between the two is connected with the fact that knowledge is only of what is true. But Newman agrees (though not with complete regularity) that there can be false certitude. Hence his position is internally inconsistent here.

However, the internal inconsistency in this case may not be very important given Newman's apologetic purpose. There is no sufficient reason for him to insist that certitude must be indefectible. Once Newman has shown, convincingly, that past mistakes do not make subsequent certainty impossible to justify, it is not of great moment whether certainties may be lost, and it becomes just a matter of the definition of certitude as contrasted with conviction. Newman, to his credit, does not ever argue 'I am certain *ergo* this is true'.

4. Fourthly, Newman's argument from conscience is unconvincing. The parallel drawn in *The Grammar of Assent* with our knowledge of the external world is based on a false phenomenalist view which most philosophers would now regard as indefensible. It is interesting that this view conflicts with that presented in the University sermons. In his later, but not his earlier, writing Newman assumes that our knowledge of material objects is indirect, a hypothesis from phenomena.

5. Conscience itself may be seen as conditioned or absolute. If conditioned, it is the result of reasoning—as it is for the Utilitarian, operating his felicific calculus. Newman is aware of this, and denounces the idea. 'We reprobate under the name of Utilitarianism, the substitution of Reason for Conscience' (U, 175). But reasoning need not be Utilitarian, and Aristotle, whose

*phronesis* Newman takes as the paradigm for illative sense, does present a theory of conscience which makes it the result of practical reasoning.

If, on the other hand, conscience is thought of not as a conclusion from reasoning, but as an absolute dictate, then the objection of J. L. Mackie tells:

> If we take conscience at its face value and accept as really valid what it asserts, we must say that there is a rational prescriptivity about certain kinds of action for doing them or for refraining from them. There is a to-be-done-ness or a not-to-be-done-ness involved in that kind of action in itself. If so, there is no need to look beyond this to any supernatural person who commands or forbids such action . . . (*The Miracle of Theism*, 104)

If the existence of God is looked on not as something perceived behind conscience, but as something to explain the origin of conscience, then of course Newman's hypothesis needs to be considered in competition with other hypotheses. One such hypothesis is the theory of Freud, which to any modern reader is brought irresistibly to mind by the passage quoted above about the delight which is received from the praise of a father.

One of the earliest readers of the *Grammar of Assent* was Gerard Manley Hopkins. He wrote to a friend

> It is perhaps heavy reading. The justice and candour and gravity and rightness of mind is what is so beautiful in all he writes but what dissatisfies me is a narrow circle of instance and quotation. . . . But he remains, nevertheless, our greatest living master of style. (*Further Letters*, ed. C. C. Abbot (London 1956), p. 58)

Hopkins offered to write a commentary to remedy the deficiencies of the book. Given the smooth sunlight brilliance of most of Newman's writing, and the dense tangled opacity of which Hopkins was master, it is not surprising that Newman rejected the suggestion with a degree of impatience. But it would have been wonderful to have had a work which combined the gifts of the two greatest masters of English of the age.

REFERENCES

G = *An Essay in Aid of a Grammar of Assent*, ed. I. T. Ker (Oxford: Clarendon Press, 1985).

U = *Sermons, chiefly on the theory of religious belief, preached before the University of Oxford*, 2nd edn. (London: Rivington, 1844).

# 8 Anselm on the Conceivability of God

Is the ontological argument valid? Professor Timothy Smiley of Cambridge once offered a succinct and trenchant argument in favour of its validity. Define the ontological argument, he said, as the best possible argument for the existence of God. Now clearly an argument for the existence of God which is valid is better than an argument for his existence which is invalid. Therefore the best possible argument for the existence of God is valid, and so the ontological argument is valid.

I shall not in this lecture be concerned with the validity of the ontological argument; I doubt if I can offer, in brief compass, anything which would improve on Professor Smiley's entertaining presentation. Instead, I shall discuss what would follow about the conceivability of God if we were to follow the line of thought of Anselm in the *Proslogion*.

Let us begin by making a contrast between the ontological argument and other forms of argument to the existence of God, such as the different versions of the cosmological argument. All such proofs start from a phenomenon, or class of phenomena, within the world, which demand explanation. They go on to show that a particular type of explanation will not lead to intellectual satisfaction, however frequently it is applied. Thus movement is not to be explained by objects in motion, nor can effects be explained ultimately by causes which are themselves in turn effects, nor can complexity be explained by beings which are themselves complex.

Proofs of the existence of God, if they are not to be mere appeals to ignorance and incomprehension, must not depend on particular features of the world which are as yet unexplained. They must depend on the necessary limits of particular types of explanation. The cosmological argument must depend on necessary, not contingent, features of the kind of cosmos to be explained. Otherwise they will be vulnerable to defeat by the progress of science. (Elsewhere I have argued that the Five Ways of Aquinas are unsuccessful forms of the cosmological argument precisely because they depend, more than at first meets the eye, on particular theories of physical explanation.)

If there is to be a successful version of the cosmological argument, it must be an argument to show that a particular type of explanation must fail to render intelligible the class of phenomena to be explained, and that intelligibility can only be found, if at all, in a being which stands outside the application of that particular paradigm of explanation. Such a being, the argument may conclude in the style of St Thomas, is what all men call God; but it is not to be taken for granted that we understand without further ado what the nature of that 'calling' may be.

The ontological argument, no less than the cosmological argument, is an argument pointing to a limit. But now the limit is not the limit of explanation, but the limit of conception itself. The premise of the ontological argument is that each of us, even the atheist, has the concept of God as that than which no greater can be conceived. From this premise, St Anselm offers to prove that God must exist in reality and not only in the mind. But it is not to be forgotten that he goes on to say that that than which no greater can be conceived cannot itself be conceived. The corollary of the ontological argument therefore appears to cut off the premise on which it rests. This is the problem with which my lecture will be concerned.

God, I have said, is not a terminus of any of the normal patterns of explanation in the world. Rather, the concept of God is invoked as a limiting case of explanation. If a proof of the existence of God is to take its start from an explanatory series in the world, it must aim to show that such a series, however prolonged, cannot arrive at a complete and intellectually satisfactory account of the phenomena to be explained. The argument must take a form similar to the demonstration that the addition of one-half to one-quarter to one-eighth . . . and so on will never exceed unity.

If we are to have a proof of the existence of God it will not suffice to say that we do not know whether some pattern of explanation in the world will succeed in explaining everything that needs explaining; we have to aim to show that it cannot. And that is indeed what the traditional proofs of God attempted to do: to show, for instance, that no explanation by one or more moving objects will suffice to explain motion, that no explanation of one contingent object by another contingent object will suffice to explain contingency.

We may look at proofs such as Aquinas' five ways as providing not so much proofs as definitions of God. God is then that which accounts for what, in the motion series, is left unexplained by previous motors in the series. God is that which accounts for that which, in the causal series, is left unexplained by the individual members of the series. God is that which accounts for what is left unexplained in the series of contingent substances which arise from each other and turn into each other. God is that which accounts for what is left unexplained in the series of complex entities composed of simpler entities.

The way in which God accounts for the unexplained is not by figuring in some further explanation. When we invoke God we do not explain the world, or any series of phenomena in the world. The mode of intelligibility which is provided by the invocation of God is something of a quite different kind. In terms of a distinction fashionable in some philosophical quarters, the introduction of the concept of God provides not explanation but understanding. The appeal to God is not based on particular failures of explanation, but upon the provable inability of a particular pattern of explanation to give an intellectually satisfying understanding of phenomena of a certain type.

Consider, for instance, the relationship of the Argument from Design to Darwinian explanation by evolution. The theist position and the evolutionary one are not competing explanations of the same fact. However successful explanation by natural selection may be in explaining the origin of particular species of life, it clearly cannot explain how there come to be such things as species at all. That is to say, it cannot explain how there came to be true breeding populations; since the existence of such populations is one of the premises on which explanations in terms of natural selection rest as their starting point.

To say this is not to say that Darwinians do not offer explanations of the origin of life; of course they do, but they are explanations of a radically different kind from explanation by natural selection. Whether God must be invoked as the author of life, or whether one of the explanations of life in terms of chance and necessity can be made intellectually satisfactory, one thing is clear: natural selection cannot explain the Origin of Species.

The nature of theistic argument here is often misunderstood by exponents of evolution. One can illustrate this by referring to the work of Richard Dawkins, whose book *The Blind Watchmaker* is one of the most lucid expositions of natural selection in the English language. On p. 141 of his book Dawkins considers the following argument offered to show the difficulties of accounting for the origin of life and the existence of the original machinery of replication:

> Cumulative selection can manufacture complexity while singlestep selection cannot. But cumulative selection cannot work unless there is some minimal machinery of replication and replicator power, and the only machinery of replication that we know seems too complicated to have come into existence by means of anything less than many generations of cumulative selection.

This argument, Dawkins says, is sometimes offered as proof of an intelligent designer, the creator of DNA and protein. He replies:

> This is a transparently feeble argument, indeed it is obviously self-defeating. Organized complexity is the thing we are having difficulty in explaining. Once we are allowed simply to postulate organized complexity, if only the organized complexity of the DNA/protein replicating engine, it is relatively easy to invoke it as a generator of yet more organized complexity. That, indeed, is what most of this book is about. But of course any God capable of intelligently designing something

as complex as the DNA/protein replicating machine must have been at least as complex and organized as that machine itself. Far more so if we suppose him *additionally* capable of such advanced functions as listening to prayers and forgiving sins. To explain the origin of the DNA/protein machine by invoking a supernatural Designer is to explain precisely nothing, for it leaves unexplained the origin of the Designer. You have to say something like 'God was always there' and if you allow yourself that kind of lazy way out, you might as well just say 'DNA was always there' or 'Life was always there', and be done with it.

A traditional theist would say that this paragraph misrepresented the notion of God in two ways. First of all, God is as much outside the series complexity/simplicity as he is outside the series mover/moved. He is not complex as a protein is; nor, for that matter, is he simple as an elementary particle is. He has neither the simplicity nor the complexity of material objects. Secondly, he is not one of a series of temporal contingents, each requiring explanation in terms of a previous state of the universe: unchanging and everlasting, he is outside the temporal series.

Because God is not a part of any of the explanatory series which he is invoked to account for—he is an unmoved mover, he is first cause only by analogy—the vocabulary and predicates of the different explanatory series are not applicable to him in any literal sense.

But when we turn from the cosmological argument to the ontological one, the vocabulary at our disposal to describe God becomes even more constrained. The ontological argument, in contrast to the cosmological argument, concerns not explanation, but conception. God, in Anselm's definition, becomes the outer limit of conception; because anything than which something greater can be conceived is not God. God is not the greatest conceivable object (and this is one reason why Professor Smiley's version of the ontological argument is only a joke). God is himself greater than can be conceived, therefore beyond the bounds of conception, and therefore literally inconceivable.

But if God is inconceivable, does that not mean that the notion of God is self-contradictory, and God a nonsensical *Unding* which cannot exist? That would be so if conceivability were mere freedom from contradiction; but there are many reasons for thinking that non-contradictoriness is only a necessary, not a sufficient condition for conceivability.

If God is not conceivable, is it not self-refuting to talk about him at all, even if only to state his inconceivability? Let us look more closely at Anselm's text to see how he handles this difficulty.

The fool says in his heart there is no God; that is to say he thinks (*cogitat*) that there is no God. On the other hand, he hears, and understands (*intelligit*) *that than which no greater can be thought.* So he thinks that that than which no greater can be thought does not exist. But how can this be since (according to chapter 3) that than which no greater can be thought cannot be thought not to exist? This is the question which is posed by chapter 4: if saying in the heart is thinking, how could the fool say in his heart what cannot be thought?

Anselm appears to reply by making a distinction between two senses of 'think' (*non uno modo cogitatur*). In one sense, I think of something if I think of a word which signifies it; in another sense I think of a thing only if I understand that which the thing is in itself. The fool can understand the words 'that than which nothing greater can be thought'; he can only deny the existence of God because he does not understand the reality which lies behind the words.

The paradox which faces Anselm in giving an account of what is going on in the fool's mind is a type of paradox which is familiar also in other areas of philosophy, such as philosophy of mathematics. Bertrand Russell gave currency to Berry's paradox, which invites us to consider the expression 'the least natural number not nameable in fewer than twenty-two syllables'. This expression names in twenty-one syllables a natural number which by definition cannot be named in fewer than twenty-two syllables. Clearly, to solve this paradox, we have at least to distinguish between different ways of *naming.*

However, the solution to the paradox which faces Anselm cannot be solved simply by distinguishing between two different ways of *thinking.* For Anselm goes on to say that not only the fool, but none of us understand what lies behind the words 'that than which nothing greater can be thought'. Let us consider a number of passages which leave the matter beyond doubt.

God lives in inaccessible light: his goodness is incomprehensible. His goodness is beyond all understanding (*bonitas quae sic omnem intellectum excedis*). (c. 9). The soul strains to see but it cannot see anything beyond what it sees except darkness—but it does not really see darkness, for there is no darkness in God, but it sees that it cannot see further because of its own darkness (c. 14). God is not only that than which no greater can be thought but is himself something greater than can be thought (*Non solum es quo maius cogitari nequit, sed es quiddam maius quam cogitari possit*). (c. 15).

There is nothing self-contradictory in saying that that than which no greater can be thought is itself too great for thought: 'that than which no greater can be thought' is not equivalent to 'the greatest possible object of thought'. I can say that my copy of the *Proslogion* is something than which nothing larger will fit into my pocket. That is true, but it does not mean that my copy of the *Proslogion* will itself fit into my pocket: in fact, it is far too big to do so.

But we may ask what sort of *cogitatio* Anselm has in mind here: the sort that deals with the words for things or the sort that deals with the essence of things? The distinction made earlier will not help here. For if God is literally ineffable, then there are not words to denote and describe him as there are words to describe and denote other things.

In the reply to Gaunilo Anselm makes no systematic distinction between being thought of and being understood: what is in the intellect and what is thought of appear to be the same. The distinction drawn there is rather between being in the intellect and being fully understood:

> Quod si dicis non intelligi et non esse in intellectu quod non penitus intelligitur: dic quia qui non potest intueri purissimam lucem solis, non videt lucem diei, quae non est nisi lux solis.

We cannot look at the sun, but we see the sun's light: we are invited to draw the parallel in the case of God.

Later in the reply a distinction is drawn between *intelligere* and *cogitare* but on a basis different from the distinction between two sorts of thought drawn in the *Proslogion*. Now the distinction seems to be on the basis that one can *intelligere* only what is true, but one can *cogitare* also what is false.

Nam et si nulla quae sunt possint intelligi non esse, omnia tamen possunt cogitari non esse, praeter id quod summe est.

Anselm's last word on the topic of the ineffability of God comes in the ninth chapter of the reply to Gaunilo.

Sed et si verum esset non posse cogitari vel intelligi illud quo maius nequit cogitari, non tamen falsum esset 'quo maius cogitari nequit' cogitari posse et intelligi. Sicut enim nihil prohibet dici ineffabile, licet illud dici non possit quod ineffabile dicitur; et quemadmodum cogitari potest non cogitabile, quamvis illud cogitari non possit cui convenit non cogitabile dici: ita cum dicitur 'quo nihil maius cogitari' valet, procul dubio quod auditur cogitari et intelligi potest, etiam si res illa cogitari non valeat aut intelligi, qua maius cogitari nequit.

Subtle as it is, this paragraph does not really solve the problem. How is it possible to know what a word means if what it means cannot even be thought about? If a thing is ineffable, what is one saying when one tries to identify the thing? The distinction between understanding words and understanding the thing which they describe can only be effective if the things in question are to some extent describable and to that extent are not ineffable.

Anselm's problem, in his own terms, seems insoluble. Does the difficulty apply to all attempts to talk about God? Not necessarily. A possible solution may be found by making a distinction between two kinds of ineffability: by exploring the suggestion that while we can speak of God, we cannot speak of him literally. God, if that is so, will be literally ineffable, but metaphorically describable.

If this is correct, then there cannot be any *science* of theology. My book *The God of the Philosophers* (Clarendon Press, 1979) was meant to be a refutation of this would-be science. The God of scholastic and rationalistic philosophy is an *Unding*, full of contradiction. Even in talking about God we must not contradict ourselves. Once we find ourselves uttering contradictory propositions, we must draw ourselves up. We can perhaps seek to show that the contradiction is only apparent; we may trace back the steps that led to the contradictory conclusion, in the hope that minor modification to one of the steps will remove the clash. The one thing we must not do is to accept contradiction cheerfully.

To say that we cannot speak literally of God is to say that the word 'God' does not belong in a language game. Literal truth is truth within a language game. Some philosophers believe that there is a special religious language game, and it is in that game that the concept of God is located. I believe, on the contrary, that there is no religious language game, and that we speak of God in metaphor. And to use metaphor is to use a word in a language game which is not its home.

However, it is not peculiar to theology that it cannot be incapsulated in a language game. If Wittgenstein is right—and after all the notion of language game is his coinage—there is no philosophical language game either: there are no truths special to philosophy. Finally, a certain kind of poetry is an attempt to express what is literally inexpressible.

I have said that theology speaks in metaphor. Theologians have preferred to say that theological language is analogical, and analogical discourse is not necessarily metaphorical. However, theological attempts to explain how non-metaphorical analogy applies to God have been, in my view, unsuccessful.

Metaphor, as has been said, is not a move in a language game. It is, in the standard case, taking a word which has a role in one language game and moving it to another. In the case of God it is taking a word which has no role in any standard language game and using it in other games. If there is such a thing as a religious language game, it is not a language game in which there is literal truth.

I know of no philosopher who has described the paradox of talking about the inconceivable godhead with such precision as the poet Arthur Hugh Clough. Consider, as an example, his poem of 1851 *Hymnos Aymnos* ('a hymn, yet not a hymn'). (*The Poems of Arthur Hugh Clough* ed. Mulhauser (Clarendon Press 1974), 311). Its first stanza begins with an invocation to the incomprehensible Godhead:

> O Thou whose image in the shrine
> Of human spirits dwells divine;
> Which from that precinct once conveyed,
> To be to outer day displayed,
> Doth vanish, part, and leave behind
> Mere blank and void of empty mind,
> Which fitful fancy seeks in vain
> With casual shapes to fill again.

If we look for God in the inmost soul we are bound to fail to find him. Attempts to give public expression to the God encountered in the soul yield only meaningless, self-contradictory utterances ('blank and void') or images unconnected with reality ('casual shapes').

The second stanza develops the theme of the impotence of human utterance to embody the divine. In the third the poet proclaims that silence—inner as well as outer—is the only response to the ineffable:

> O thou, in that mysterious shrine
> Enthroned, as we must say, divine!
> I will not frame one thought of what
> Thou mayest either be or not.
> I will not prate of 'thus' and 'so'
> And be profane with 'yes' and 'no'.
> Enough that in our soul and heart
> Thou, whatso'er thou may'st be, art.

The agnosticism is radical: the *via negativa* is rejected as firmly as the *via positiva*. Not only can we not say of God what he is, we are equally impotent to say what he is not. The possibility, therefore, cannot be ruled out that one or other of the revelations claimed by others may after all be true:

> Unseen, secure in that high shrine
> Acknowledged present and divine
> I will not ask some upper air,
> Some future day, to place thee there;
> Nor say, nor yet deny, Such men
> Or women saw thee thus and then:
> Thy name was such, and there or here
> To him or her thou didst appear.

In the final stanza the agnosticism is pushed even further. Perhaps there is no way in which God dwells—even ineffably—as an object of the inner vision of the soul. Perhaps we should reconcile ourselves to the idea that God is not to be found at all by human minds. But even that does not take off all possibility of prayer.

> Do only thou in that dim shrine,
> Unknown or known, remain, divine;
> There, or if not, at least in eyes
> That scan the fact that round them lies.
> The hand to sway, the judgment guide,
> In sight and sense, thyself divide:
> Be thou but there,—in soul and heart,
> I will not ask to feel thou art.

The soul reconciled to the truth that there can be no analogue of seeing or feeling God, that nothing can be meaningfully said about him, can yet address Him and pray to be illuminated by his power and be the instrument of his action. But does not this presume that God can after all be described: at least as a powerful agent who can hear our prayers? No, the prayer need not assume the truth of that; only its *possibility* is needed to make sense of agnostic prayer.

The ineffability of God, thus movingly expressed by Clough, has been frequently proclaimed by the most orthodox thinkers including, as we have seen, St Anselm. But from the ineffability of God orthodox believers have never drawn the conclusion that it is profane to use words to describe and invoke him. Rather, they have said, with Saint Augustine, *vae tacentibus de te*—woe to those who are silent about thee.

Some religious thinkers have attempted to show that coherent literal description of God is after all possible; others have simply claimed that there can be worse things than talking nonsense. Perhaps that is what lies behind Augustine's *vae tacentibus*. We may aim at a rational worship, and yet get no further than the babble of infants or the glossolaly of the possessed.

In the present century no man surpassed Wittgenstein in the devotion of sharp intelligence to the demarcation of the boundary between sense and nonsense. Wittgenstein finished the masterpiece of his youth with the words 'Wovon man nicht sprechen kann, darüber muss man schweigen': whereof one cannot speak, thereof one must be silent. But within ten years he was putting forth his own gloss on Augustine's *Vae tacentibus*.

'Was, du Mistviech, du willst keinen Unsinn reden? Rede nur einen Unsinn, es macht nichts' (F. Waismann and B. F. McGuinness, *Ludwig Wittgenstein und der Wiener Kreis* (Blackwell 1967), 69). Which we may paraphrase thus: so you don't want to talk nonsense, you cowpat? go on, talk nonsense, it won't do you any harm.

# REFERENCES

ANSCOMBE, G. E. M., 'On Believing Someone' in C. F. Delaney (ed.), *Rationality and Religious Belief* (Notre Dame, 1979).

AYER, A. J., *Language, Truth and Logic* (London, 1946).

CLIFFORD, W. K., 'The Ethics of Belief' in *Lectures and Essays* (London, 1879).

CLOUGH, ARTHUR HUGH, *Poems*, ed. F. Mulhauser (Oxford, 1974).

DAWKINS, RICHARD, *The Blind Watchmaker* (London, 1986).

FARRER, AUSTIN, *Finite and Infinite* (London, 1940).

FLEW, ANTONY, *God and Philosophy* (London, 1966).

——, MACINTYRE, ALASDAIR, *New Essays in Philosophical Theology* (London, 1955).

KENNY, ANTHONY, *The Five Ways* (London, 1969).

——, *The God of the Philosophers* (Oxford, 1979).

MACKIE, JOHN, *The Miracle of Theism* (Oxford, 1982).

NEWMAN, JOHN HENRY, *An Essay in Aid of a Grammar of Assent*, ed. I. T. Ker (Oxford, 1985).

——, *Sermons, chiefly on the theory of religious belief, preached before the University of Oxford* (London, 1844).

PLANTINGA, ALVIN, 'Is Belief in God Rational?' in C. F. Delaney (ed.) *Rationality and Religious Belief* (Notre Dame, 1979).

——, 'Is Belief in God Properly Basic?' *Nous* (1981).

——, 'The Reformed Objection to Natural Theology' *Christian Scholar's Review* (1982).

PRICE, H. H., *Belief* (London, 1969).

QUINE, W. V. and ULLIAN, J. S., *The Web of Belief* (New York, 1978).

WITTGENSTEIN, L., *On Certainty* (Oxford, 1969).

# INDEX

Index

## MORE OXFORD PAPERBACKS

Details of a selection of other Oxford Paperbacks follow. A complete list of Oxford Paperbacks, including The World's Classics, Twentieth-Century Classics, OPUS, Past Masters, Oxford Authors, Oxford Shakespeare, and Oxford Paperback Reference, is available in the UK from the General Publicity Department, Oxford University Press (RS), Walton Street, Oxford, OX2 6DP.

In the USA, complete lists are available from the Paperbacks Marketing Manager, Oxford University Press, 200 Madison Avenue, New York, NY 10016.

Oxford Paperbacks are available from all good bookshops. In case of difficulty, customers in the UK can order direct from Oxford University Press Bookshop, 116 High Street, Oxford, Freepost, OX1 4BR, enclosing full payment. Please add 10 per cent of the published price for postage and packing.

## PAST MASTERS

*General Editor: Keith Thomas*

*Past Masters* is a series of authoritative studies that introduce students and general readers alike to the thought of leading intellectual figures of the past whose ideas still influence many aspects of modern life.

'This Oxford University Press series continues on its encyclopaedic way . . . One begins to wonder whether any intelligent person can afford not to possess the whole series.' *Expository Times*

## KIERKEGAARD

*Patrick Gardiner*

Søren Kierkegaard (1813–55), one of the most original thinkers of the nineteenth century, wrote widely on religious, philosophical, and literary themes. But his idiosyncratic manner of presenting some of his leading ideas initially obscured their fundamental import.

This book shows how Kierkegaard developed his views in emphatic opposition to prevailing opinions, including certain metaphysical claims about the relation of thought to existence. It describes his reaction to the ethical and religious theories of Kant and Hegel, and it also contrasts his position with doctrines currently being advanced by men like Feuerbach and Marx. Kierkegaard's seminal diagnosis of the human condition, which emphasizes the significance of individual choice, has arguably been his most striking philosophical legacy, particularly for the growth of existentialism. Both that and his arresting but paradoxical conception of religious belief are critically discussed, Patrick Gardiner concluding this lucid introduction by indicating salient ways in which they have impinged on contemporary thought.

Also available in Past Masters:

*Disraeli* John Vincent
*Freud* Anthony Storr
*Hume* A. J. Ayer
*Augustine* Henry Chadwick

# HISTORY IN OXFORD PAPERBACKS

Oxford Paperbacks' superb history list offers books on a wide range of topics from ancient to modern times, whether general period studies or assessments of particular events, movements, or personalities.

## THE STRUGGLE FOR THE MASTERY OF EUROPE 1848–1918

*A. J. P. Taylor*

The fall of Metternich in the revolutions of 1848 heralded an era of unprecedented nationalism in Europe, culminating in the collapse of the Hapsburg, Romanov, and Hohenzollern dynasties at the end of the First World War. In the intervening seventy years the boundaries of Europe changed dramatically from those established at Vienna in 1815. Cavour championed the cause of *Risorgimento* in Italy; Bismarck's three wars brought about the unification of Germany; Serbia and Bulgaria gained their independence courtesy of the decline of Turkey—'the sick man of Europe'; while the great powers scrambled for places in the sun in Africa. However, with America's entry into the war and President Wilson's adherence to idealistic internationalist principles, Europe ceased to be the centre of the world, although its problems, still primarily revolving around nationalist aspirations, were to smash the Treaty of Versailles and plunge the world into war once more.

A. J. P. Taylor has drawn the material for his account of this turbulent period from the many volumes of diplomatic documents which have been published in the five major European languages. By using vivid language and forceful characterization, he has produced a book that is as much a work of literature as a contribution to scientific history.

'One of the glories of twentieth-century writing.' *Observer*

Also in Oxford Paperbacks:

*Portrait of an Age: Victorian England* G. M. Young
*Germany 1866–1945* Gorden A. Craig
*The Russian Revolution 1917–1932* Sheila Fitzpatrick
*France 1848–1945* Theodore Zeldin